Mathematica

MATHEMATICA

A SECRET WORLD OF INTUITION AND CURIOSITY

• • •

DAVID BESSIS

TRANSLATED BY KEVIN FREY

Yale
UNIVERSITY PRESS
New Haven and London

Published with assistance from the Centre national du livre.

Published with assistance from the foundation established in memory of Philip Hamilton McMillan of the Class of 1894, Yale College.

Originally published in French as *Mathematica: Une aventure au coeur de nous-mêmes* by Éditions du Seuil in 2022.

Quotations from *Harvests and Sowings (Récoltes et semailles)* by Alexander Grothendieck, including the epigraph, appear in translation by permission of The MIT Press, with translations by Kevin Frey.

Yale University Press books may be purchased in quantity for educational, business, or promotional use. For information, please e-mail sales.press@yale.edu (U.S. office) or sales@yaleup.co.uk (U.K. office).

Set in Adobe Garamond type by Integrated Publishing Solutions.
Printed in Great Britain by TJ Books Limited, Padstow, Cornwall

Library of Congress Control Number: 2023942848
ISBN 978-0-300-27088-4 (hardcover : alk. paper)

A catalogue record for this book is available from the British Library.

10 9 8 7 6 5 4 3 2 1

Lending an ear to the Dreamer within us is communicating with ourselves, in spite of the powerful barriers that aim, at whatever cost, to forbid us from doing so.

Alexander Grothendieck

Contents

Mathematica

1

Three Secrets

The aim of this book is to change the way you see the world.

It is grounded in my personal journey, a long adventure that has physically transformed me and endowed me with magical powers. But this journey is not mine alone. It is a collective journey, one of the most ancient and powerful ever. Begun at the dawn of time by a handful of human beings, it continues to this very day to transform our civilization, language, and thought.

How many among us have felt mathematics live and grow within ourselves? I don't know. I just know that we are a tiny minority and that our story is as yet misunderstood.

Mathematics has the reputation of being inaccessible. You have to be one of the elite, to have received a special gift. The greatest mathematicians have written that this isn't so. What they accomplished, as we shall see, they claim to have accomplished through ordinary human means, their curiosity and imagination, their doubts and weaknesses.

No one wanted to believe them. Perhaps they didn't know how to tell their story in simple enough language. Or perhaps they underestimated the force of the myth that they called into question, one of the last great human myths: the myth of intelligence.

Mathematics shapes our world. It is an instrument of power and domination. But for those who live it, math is above all an inner experience, a sensual and spiritual quest.

This experience has little to do with what we're taught in school. In certain ways, it's a form of clairvoyance, of psychic thought. In other ways, it's a resurgence of the mysterious phenomenon that in our childhood allows us to learn how to speak.

Understanding math is to travel along a secret path that brings us back to the mental plasticity we had as children. It's to discover how to reactivate and domesticate that plasticity. It's to choose to bring it back to life. This intellectual path is surprisingly close to that which we take in our everyday lives. But its entrance is hidden, concealed behind our habits, behind our fears and inhibitions. I would like to help you find this path.

There Has to Be Something Else

"I have no special talent. I am only passionately curious."

When I was fifteen, I hated this quote from Einstein. To me it sounded phony, insincere, like a supermodel saying that what really counts is inner beauty. Do we really need to hear this stuff?

The main message of this book, however, is to take Einstein's words seriously.

When you think about it, it is surprising that we have such a hard time taking him seriously. Einstein doesn't have the reputation of being a complete idiot or a compulsive liar. If you ask people in the street, they'll say that his theory of relativity is one of the great contributions to human thought. What Einstein said and wrote, therefore, merits our attention.

But when he suggests that his creativity might be accessible to others, that it simply follows from a slightly different approach that anyone might take, we find it hard to believe him. The poor old guy doesn't know what he's saying. Or worse, it's false modesty, and he's saying it just to show off.

The problem is that once you refuse to take Einstein's statement seriously, you cut off the conversation—a conversation that deserves to be pursued.

Einstein's statement is objectively intriguing but it doesn't really say much. Let's assume that it is correct. What are we supposed to do with it? How can it help us? Without any concrete details or practical advice, it's difficult to learn anything from it.

It's kind of surprising that no one had the presence of mind to respond, "Albert, what you just said is really interesting, but we'd like to know more. Could you explain it to us? We want to know the secret details, to learn how you really do it. Do you want to go get a coffee? Or maybe go for a nice walk in the woods? Come on and tell it all; we have loads of questions!"

The first questions I would have liked to ask are pretty inane:

1. Albert, where does your curiosity come from?

I don't know many people curious enough to shut themselves in a room and meditate upon problems in theoretical physics. But I do know a few, and they all say the same thing: that if they shut themselves in a room and study problems in theoretical physics, it's of course in part scientific ambition, but mostly because they get real pleasure from it.

So the question becomes: *Albert, how do you derive pleasure from studying physics*?

2. How do you keep from getting discouraged?

To be *passionately* curious means to have the ability to be interested in things with an unwavering commitment, with an intensity and tenacity that never fails. Einstein clearly found a secret means of not giving up where others had faltered. What was his secret?

Doing research in pure mathematics has taught me one essential thing: that when you shut yourself in a room with a difficult problem, you have only one wish: to get out of there as fast as you can.

It's simply terrifying to reach the limits of your intelligence, to push on in vain, to struggle for months, to feel too stupid to understand and have no idea how to pull through.

Einstein found a way to tame his fears and resist the impulse to flee. How did he do it?

3. When you're alone in a room with a problem, what exactly is going on?

Or, to be more explicit: what did Einstein do with the problem? How did he get his hands on it? What was he playing around with?

It might seem silly to use such trivial language, but let's be honest: what we really want to know are the juicy details. We want to know what *really* went on inside Einstein's head. We want to know how he actually did it. We want to know Einstein's technique, his secret magic that worked every time.

We know that intellectual creativity isn't just a question of how much work you do. We know there has to be something else, a secret ingredient, something mysterious that's never even mentioned at school.

If Einstein had taken the time to teach us his method for achieving great scientific discoveries, his contribution to humanity would have greatly surpassed his work in physics. As the saying goes, give someone a fish and they're hungry again tomorrow; teach them how to fish and they'll have food for a lifetime.

But this discussion never took place. It never will take place. Albert Einstein died on April 18, 1955, at the University Medical Center of Princeton. The doctor who performed the autopsy was himself so eager to discover the secret of Einstein's genius that, without the consent of the family, he removed the brain and sliced it into thousands of pieces.

He didn't learn much from it.

The Method

This problem, however, goes far beyond Einstein. It's gone on for centuries. It concerns our false beliefs, our misconceptions about intelligence and creativity, and the extent to which these false beliefs limit us.

The most difficult thing about understanding Einstein's work is mathematical formalism. It's also what caused Einstein himself the greatest trouble, as he admitted to a high school student who asked him for advice: "Do not worry about your difficulties in Mathematics. I can assure you mine are still greater."

Four hundred years ago the greatest mathematician of the time talked about his life in a book that has since become famous. His message is perfectly clear from the outset. It can be summed up as follows: "I am not any more intelligent than the others. I simply had the chance to discover a magical method that allowed me to become better than anyone else. I will tell you how I did it."

The same knee-jerk reaction that makes it difficult to take Einstein's words seriously also keeps us from understanding what this mathematician (René Descartes) is trying to tell us, and keeps us from placing his book (*Discourse on Method*) where it belongs: in the self-improvement section.

The consensus is that there isn't any method for becoming a great mathematician, any more than there is for losing weight by drinking milkshakes or getting rich by working from home for two hours a week.

Little does it matter that Descartes is telling us the exact opposite.

Three False Beliefs

We'll talk more about *Discourse on Method* in chapter 14, but in order to understand what Einstein and Descartes are trying to tell us, we have to begin by ridding ourselves of three common beliefs about mathematics:

1. In order to do mathematics, you need to think logically.
2. A few of us are naturally at ease with numbers and a few others naturally have a good geometric intuition. Unfortunately, the great majority of people understand nothing about math, and can't do anything to change that.
3. Great mathematicians are born with a brain fundamentally different from ours.

We may as well be clear about the first one: no, mathematicians don't think logically. It is in fact utterly impossible to think logically. Logic doesn't help at all with thinking. We shall see later on what it is used for.

The second fallacy is truly toxic. It has the power to make us hopelessly inhibited. It has actually succeeded at convincing most of humanity that math is a strange and dangerous territory. For each of us, including the most "gifted," it imposes an unsurpassable limit, that of the mathematical intuition everyone is "naturally" endowed with.

The third misconception is a simple variation on the same theme: to be like Einstein or Descartes, you have to be born that way; you can't get there by trying. And when Einstein or Descartes tell us differently, they're just making fun of us.

This vision that we're incapable of *becoming* good at math is false, but it derives from an essential truth: the magic power of mathematicians isn't logic but intuition.

Official Math vs. Secret Math

Einstein liked to talk about the importance of intuition in his discoveries. "I believe in intuition and inspiration," he said, and he was being quite serious when he said it. As for mathematicians, they know quite well that there exist two different kinds of math.

Official math can be found in textbooks, where it is presented in a logical and structured manner, in an esoteric language that relies on indecipherable symbols.

Secret math, also known as *mathematical intuition,* can be found in the heads of mathematicians. It consists of mental representations and abstract sensations, often visual, that are for them quite obvious, and that give them a great deal of pleasure. But when it comes to sharing these sensations with the rest of the world, mathematicians are often at a loss. What had seemed so evident to them is suddenly less so.

To transcribe their ideas, mathematicians have had to invent that esoteric language and those indecipherable symbols, just as musicians had to invent a complex musical notation in order to transcribe their compositions. Except that musicians have one enormous practical advantage: they only have to have their music played for everyone to immediately understand what it's about, without needing to decipher the written score.

Mathematicians don't have this option and it's a huge problem for them. In their minds, the ideas are luminous, simple, and powerful. On paper, they become stunted and sad. The mathematicians' curse is that they can only *play* math in their own heads.

If you taught children music by giving them the written scores for Mozart or Michael Jackson to decipher without their ever having heard it played, music would be as universally hated as math.

Intuition is the soul of mathematics. Without intuition, math becomes meaningless. But you mustn't conclude from this that if you don't understand anything about math, then there's nothing you can do to change that.

The mistake is in believing that our mathematical intuition is a static given, an insurmountable limit. The intuition that we have of mathematical objects isn't innate. It's not fixed. We can build it up, make it stronger day by day, as long as we follow the right method.

Mathematicians are well aware that official math doesn't tell all the story. They know that the real goal is to *understand* what's in the books, to *see* it, to *feel* it. What mathematicians do on a daily basis is to develop their intuition, to make it richer, clearer, more powerful. Even more so than the publications and official works, mathematicians' intuition is their masterpiece, their lifetime accomplishment.

This extraordinary art of seeing the unseen, of feeling what can't be felt, of understanding at the deepest level, to the point where it becomes self-evident, what 99.9999 percent of humanity deems grotesquely abstract and utterly unintelligible—this is mathematicians' great art and their true secret. Only those who have mastered this art know how far it can lead.

But how do they do it? That's the subject of this book.

Three Secrets of Mathematicians

1. *Doing math is a physical activity.* To become capable of understanding what you don't yet understand, you have to perform specific actions in your head. These actions are invisible yet indispensable. They aim at expanding your intuition and developing new, deeper, and more powerful mental representations. In the short run this activity can be exhausting, but in the long run it makes you incredibly stronger. Learning how to do math is learning how to make use of your own body. It's like learning how to walk, swim, dance, or ride a bike. These unseen actions aren't innate, but we all have the ability to learn them.

2. *There's a way to become good at math.* This method is never taught in school. It doesn't resemble any academic method and goes against the traditional tenets of education. It tries to make things easier rather than more difficult. You can compare it to meditation, yoga, rock climbing, or martial arts. It includes techniques to overcome our fears, conquer our flight reflex in the face of the unknown, and find pleasure in being contradicted. The method's exact scope is actually broader than math. It's a universal method for reprogramming our intuition and, in that sense, it's a method for becoming more intelligent.

3. *The brains of great mathematicians work the same way as ours.* There's no doubt that natural aptitude in math, like natural aptitude in any other physical activity, isn't equally distributed among individuals. But these biological differences play a far lesser role than most people assume.

The latter point is undoubtedly controversial. A striking aspect of mathematics is the regular occurrence of incredibly talented individuals who appear out of nowhere and, from a young age, demonstrate abilities vastly superior to that of their peers. At the other end of the spectrum, many people struggle with high school math or even primary school math. In the absence of a better explanation, it is natural enough to attribute this extreme level of inequality to "innate" talent.

But the competence gap that requires explaining is, in fact, too extreme for genetics. Human beings do exhibit innate biological differences, but overall we are a fairly homogeneous species. People differ in height, muscular strength, cardiac output, and lung capacity, and part of this variability can be traced to genetic factors. Yet those differences never encompass multiple orders of magnitude.

To use a metaphor that we'll develop in the next chapter, math is so unequal that it's as if some people could run the one-hundred-meter dash in under a second, while the majority wouldn't make it in a week. While it's conceivable that some people may genetically be endowed with a neuronal metabolism that is more efficient and powerful, making them, let's say, *twice* as capable in math, or why not *ten times* as capable in math, it's hard to believe that genes alone could explain such an absurd level of inequality.

Here is a simpler and much more credible explanation. Developing good mental habits, adopting the right psychological attitude, can make you *a billion times* better at math. But the method for becoming good at math has never been taught in schools. You can reach it only by accident. You're left to discover, by yourself and by chance, snippets of the method. Most people end up not discovering anything, because certain essential points of the method are surprising and counterintuitive. It's very easy to overlook them.

The brains of great mathematicians work in the same way as ours. But their personal history, the way in which they developed their own experience of the world around them, gave them the opportunity to familiarize themselves with the method from early childhood. They found their own way, without following a set path and without knowing what they were doing, by dumb luck.

An Oral Tradition

Mathematics is often defined as the study of numbers, shapes, and other types of abstract structures. Alternatively, some define it through its formal aspects: the symbols and formulas, the axioms and theorems, the systematic use of logical deduction. But a few definitions are careful enough to add this curious caveat: *no one really knows how to define mathematics.*

For example, as I'm writing these lines, the Wikipedia page under "Mathematics" states that "there is no general consensus among mathematicians about a common definition for their academic discipline."

A key message of this book, however, is that there is a latent consensus among mathematicians about what it means to *do* math and what it *feels* like. The entire book can be read as an attempt to document this unexpressed consensus and "leak it" to the general public.

If this consensus were to be turned into a definition, it wouldn't characterize math in terms of what it studies, but as a human activity of a particular nature. Meanwhile, mathematics remains the only academic discipline that is universally taught without anyone having agreed on what it's supposed to be, which leads to some truly bizarre consequences.

For example, many mathematicians have spoken about their feeling of having been self-taught. In light of the prominent role of math in the curriculum, this is a startling paradox. Of course they're not really self-taught, since they learned a lot in school. But they are self-taught in the sense that the most important things were not taught at school.

I'm one of these paradoxical autodidacts. I learned the basics of official math at school. At the same time, without anyone teaching me, I discovered the rudiments of secret math.

For a long time, I wasn't aware of the relationship between the invisible actions I was performing in my head and being good at math. It was simply a habit I'd developed, a particular way of using my imagination.

I'll talk later about the exercises in imagination that I began to do from childhood on. At first, it was nothing more than innocent games. For example, I had fun walking around the room with my eyes closed while trying to remember the layout of the furniture. What did this have to do with what I was learning at school?

I wasn't even particularly good. I often ran into the walls. I never imagined that this game, and other increasingly difficult ones, would allow me to develop, starting out at the same level as everyone else, a particularly powerful geometric intuition.

This geometric intuition has been the secret weapon in my mathematical career. I began to see things that no one else had seen, to solve problems that no one else had been able to solve.

It was only much later, talking with other mathematicians and reading the stories of famous mathematicians, that I found out my experience was not at all unique.

While the official knowledge has been transcribed in textbooks, the secret art of mathematicians has remained an oral tradition passed down from generation to generation. It reveals what no one dares write down in books because it doesn't seem serious enough, because it's not science, and because it resembles self-improvement too much.

This story deserves to be told with simple and easily understandable language, because it concerns all of us, whether you're bad at math or a math whiz, young or old, artistically or scientifically minded. It talks of our strengths rather than our weaknesses, of our hidden talents and what we can accomplish.

Math is an inner adventure, secret and silent. But it's a universal adventure, a journey into the depths of human intelligence, consciousness, and language.

In private conversations between mathematicians, when there's no one else around to overhear them, they can finally talk about how they really see things.

Yes, math is scary. Yes, it can seem incomprehensible. Yes, it feels like you'll never understand it. And yet, there's a way to get there.

2

The Right Side of the Spoon

My son Aram is one year old and he's learning how to eat with a spoon. And, truth be told, it's a disaster. In two minutes flat he manages to get his food all over the place—on the walls, in his hair, everywhere.

I try to help him. I half-fill his spoon and give it to him. But he grabs the wrong end, the end with the food in it. I tell him he should grab the other end, the handle, and I show him how to do it. But he always insists on grabbing the end with the food. It makes sense, after all, because it's the food that he wants. Except that isn't how it's done.

Yet I'm not really worried about it. He'll get there. Everyone ends up understanding which end of the spoon to grab. I've never heard anyone say: "Spoons aren't really my thing. I've never seen the point. They really get on my nerves, so I just don't use 'em."

Humans don't have any problems with spoons. Nobody hates spoons. Spoons don't hate anyone. They're among the first tools we encounter in our lives; we use them every day and we use them forever. At first, they're mysterious and strange. Then they become familiar. Pretty soon, we're using them without thinking about it, like our own hands. And in a way, they're not that different from our own hands: our brain has internalized spoons, their uses and possibilities. They've become extensions of our own bodies.

When you know how to eat with a spoon, it's easy as pie. When you don't, it's immensely hard. We've learned to use a spoon so well that we've forgotten that we had to learn it. We've forgotten that, at first, it was far from easy.

The complexity of this action only become obvious once you watch a baby trying to do it. It requires excellent hand-eye coordination. Simply grabbing the spoon and holding it correctly takes a bit of work. Not to mention that the right way to hold a spoon depends on what you're eating with it.

It's already been fifty years since we've landed on the moon, but we're just beginning to learn how to program robots capable of eating baby food with a spoon. And let's not even talk about kiwi, which is a whole different ball game.

The Serious Stuff

Spoons are only the beginning. Then the serious stuff starts. We learn how to put on our shoes and take them off. We learn how to brush our teeth and cut our fingernails. We learn how to ride a bike and roller-skate. We learn how to peel onions and make coffee. We learn how to play video games and sew on a button. We learn how to drive and decalcify the coffeemaker. Sometimes it's a little hard at first, but pretty soon we get the hang of it.

Like the spoon or the bicycle, our tools end up becoming extensions of our selves. We use them without thinking. They transform us. They augment us. They make us what we are. Without our tools, we really don't amount to much.

Language is the most difficult thing of all to learn. It's an incredibly long, frighteningly difficult process. At eighteen months old, hardly anything we babble is intelligible. And yet we keep on trying all day long.

It's enough to get you down, but we never give it up. No one ever says: "Language, that's really not my thing. It just ain't worth it. Frankly, it's a pain."

No parent ever says: "Jane's just so cute with her pacifier that it breaks our heart to make her do anything so hard. So we've decided not to speak to her."

Language isn't an option. It's not just something for the upper class, the rich, the geniuses. It's for everyone.

If you really wanted to mark a date when we first became human, you could pick the day when our ancestors decided to give language to everyone. Well before organized religion or codified laws, we chose to follow this implicit rule: "Thou shalt teach thy children language."

A Radical Success

More recently, in the past two hundred years or so, we made a new fundamental decision: to teach everyone how to read and write. This decision is so foundational that it's become difficult to imagine what our world would look like without it, if only a small minority of the population, as was the case before, was able to read.

In ancient Egypt, with the use of hieroglyphics, the art of writing was akin to magic. Scribes were a hereditary caste, passing down their secrets from generation to generation. In medieval Europe, writing was a vocation. Young men became monks, shut themselves off from the world, and devoted their existence to copying manuscripts.

What did the peasants think of all that? Did they believe that reading and writing required a special talent, a particular form of intelligence that they didn't have? Did they find being excluded from written language unfair and frustrating? Or did they simply tell themselves they didn't have the time, money, or desire, and that in any case there wasn't anything for them to read?

Today, no one thinks that reading and writing require a special

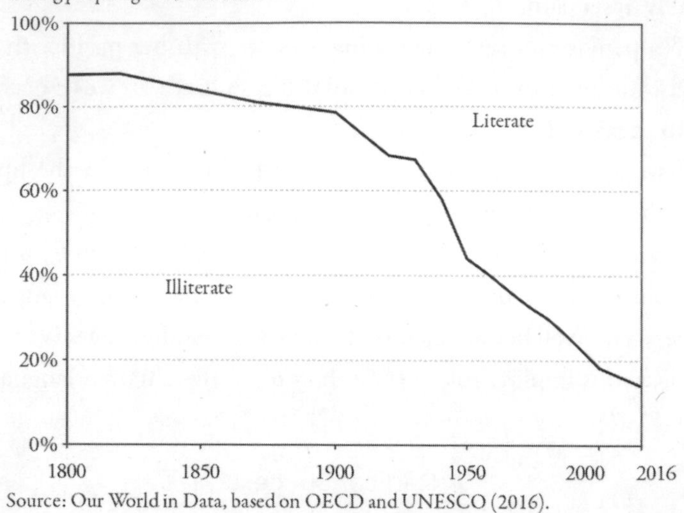

Source: Our World in Data, based on OECD and UNESCO (2016).

gift. And no one believes it's not of any use. With rare exceptions, all forms of government, whatever their religious or ideological beliefs, make primary education an absolute priority.

The radical project of global literacy has been a huge success. Illiteracy hasn't disappeared, of course, but it's become much more rare. In a few generations, humanity was able to accomplish a global program of cognitive transformation without equal in history.

A Total Disaster

At the same time that the great campaign of global literacy was being undertaken, another radical decision was made: teach everyone the basics of math. Today, in elementary and high schools around the world, more than a billion children study math.

And it's a total disaster.

Hundreds of millions of children suffer in silence. They feel like they don't understand anything, and flip-flop between utter detachment (they see absolutely nothing useful in studying math) and the humiliating sensation of simply not being smart enough.

If you ask American teenagers what's the most difficult subject, math ranks at the top of the list, with 37 percent. It's also, by far, the most hated. But if you ask what's their favorite subject, math is again in first place, with 23 percent. For some students, it's the easiest subject.

We're all aware of this strange phenomenon. It's part of the furniture, and we've come to view it as normal. We find it normal that there are people who love math and find it very easy, and others who hate it and find it incomprehensible, with practically no one in between.

We find the situation so normal that attitudes toward math have become part of our cultural stereotypes: the nerd with bad skin and glasses who loves it, the cool girl fashionista who hates it, the rebellious high school dropout who couldn't care less about it.

These stereotypes are stupid and insulting. I know dropouts who have become great mathematicians. A high school girl has the right to be pretty and popular, and still love math. She also has the right to become a great mathematician.

We've grown accustomed to it, but the situation isn't at all normal. It's really rather strange. It shouldn't have happened like that.

Just compare math learning with other basic types of learning. Would it be normal for teens to think it was cool not to know how to read? To figure that those who could read well, without having to sound out each letter, necessarily are some kind of weirdos?

Would it be okay for half of a graduating high school class not to know how to eat with a spoon? Or tie their own shoelaces?

Solving high school math problems should be as easy as tying your shoelaces, and if that's not the case, then there's something wrong with the way we're teaching math.

Two Hypotheses

To explain why some people are good at math and others aren't, two hypotheses are usually floated.

The first is that it's simply a question of motivation. People aren't good at math because they don't like it, and they don't like it because they don't see how it's useful in their day-to-day lives. But do people really think that history, for example, is useful in their everyday lives? That doesn't make it any less intelligible, and history classes don't throw people into a state of panic. You've never seen students start crying because they don't understand what a war or a revolution is.

In fact, students who aren't good at math understand very well that it's useful for something, if only to succeed at school and get into a good university. They're not stupid. They know very well that being bad at math means they won't be able to go into any number of professions, including some of the most prestigious and best paid. Maybe they don't understand why math is so important, but they know that it is. They feel excluded, which gives them an excellent reason for hating it.

The second hypothesis is just plain mean. It supposes that there's a mysterious type of intelligence, *mathematical intelligence,* that's unequally distributed amongst the population. This explanation is based on biology, postulating that there's some kind of math gene. Those who are good at math are simply born that way, and the others are out of luck.

That this idea is so widespread is somewhat surprising in itself. We should have learned by now to be wary of these kinds of ideas. There was a time when people believed that certain races were naturally made for working in the fields, while others were made for owning the plantation. More recently, it was said that women were incapable of flying fighter jets. Today, these ideas have been discredited.

If you're still skeptical, you'll see in the next chapter that you have all the intellectual abilities necessary to be good at math.

Biological inequality between individuals does exist, but it's nothing like the examples above. It's something more like this. Imagine you had a senior high school class run a hundred meters. The vast majority would be able to complete the race. Some would need eleven seconds, others thirteen or eighteen. And maybe it would take a few of them thirty seconds to run that far.

To explain these gaps, genetics is just one factor, alongside motivation, nutrition, lifestyle, and how much training the runners did. We're not all genetically similar when it comes to running a track race. But for a hundred-meter race, these genetic factors typically account for only a couple of seconds at most.

Now imagine a senior class running a hundred-meter race in which some finish in one second, but after a week more than half haven't even made it. This is about the kind of gap that you see in math skills at the end of high school.

You go looking for the students who haven't made it. Some are sitting at the starting line. They tell you that the hundred-meter is the worst thing in the world. They have no idea what use it could be in their everyday lives, and they think that the gym coach is a sadist.

Would you seriously conclude that the explanation is genetic?

I want to convince you that the only possible explanation is that it's all a giant misunderstanding. People aren't good at math because no one has taken the time to give them clear instructions. No one has told them that math is a physical activity. No one has told them that, in math, there aren't things to learn, but things to do.

They're grabbing the wrong end of the spoon because no one has ever told them that there was a right end to it.

The words spoken by the math teacher aren't the kind of things we really need to retain. They're simply instructions and indicators for the unseen actions that each of us has to do secretly inside our own head.

Studying math the same way that you study history or biology is useless. You might as well take careful notes during a yoga class so that you don't forget anything. If you don't practice any breathing exercises, it's worth nothing at all.

3
The Power of Thought

Imagine a circle, perfectly round, without any defect. Any old circle. Got it?

In real life, perfect circles don't exist. When you draw a circle on paper, there are always slight defects. Nothing is ever perfectly round—not bike tires, not the sun, not ripples on the water.

But that certainly doesn't stop you from understanding what I'm talking about, or being able to imagine a perfect circle.

Not only can you imagine it, you can literally see it. You can move it around in your mind. You can make it larger or smaller. You can do whatever you want with it.

This ability to see things that don't exist in real life, to feel that they're there, right in front of you, to move them around in your head as easily as if you could touch them—this is one of your magic powers.

It's the starting point on the road that will lead you to really understanding mathematics.

Our Incredible Capacity for Abstraction

A perfect circle is a mathematical abstraction. If circles seem like familiar objects, it's because you, like all other humans, have a natural capacity for mathematical abstraction.

But your capacity for abstraction isn't limited to math.

Whether you want to or not, you spend much of your time viewing the world abstractly. It's a physiological characteristic of your body. Your brain is a machine for creating abstractions from your sensory inputs and mentally manipulating them, just as your lungs are machines for extracting oxygen from the air and transferring it to your blood.

How is it possible? That will be the subject of chapter 19, where we'll see how the structure of our brain *naturally* allows us to create and manipulate abstractions.

Until then, and even if you don't fully understand how such a miracle is possible, you have to admit that you're able to visualize a circle.

Our Incredible Capacity for Reason

Can a straight line intersect a circle at three points?

Take your time. It's not a trap. Just try to decide for yourself. Try to imagine all the ways a straight line can intersect a circle and see if there's any way it can do so at three points.

No, a straight line can't intersect a circle at three points.

The answer seems obvious? That's because, like all human beings, you have an incredible capacity for reason. Not only are you capable of imagining abstract objects like straight lines and circles, but you're able to ask yourself abstract questions about these objects and manipulate them in your head until you find the answers.

The answer seems obvious to you, but what would you do if someone told you they didn't understand?

You'd probably want to start by saying "You can see that . . . ," but that won't work. If a person doesn't understand, it's because they can't see circles and straight lines as clearly as you do. Explaining math is getting others to see things they've never seen before.

The reasoning you used to find the answer is intuitive and visual. In your head, it's a kind of cartoon where the main characters are a circle and a straight line. This type of reasoning is very effective but difficult to translate into words. Words can never fully express all the subtleties you see in your head.

By studying math, you can learn how to translate your visual intuition into rigorous proofs. It will never be a perfect translation. It takes a lot of words to express a simple intuition. It all seems so clear in your head. But once you start to write it down, it seems technical and complicated.

Our Incredible Intuition

You're the only person capable of seeing what's in your head. Even if it's painful, it's only by making the effort of rigorously translating your vision into words and symbols that you can share it with others. And it's also the only way to make sure that your intuition is right.

Because sometimes your intuition is wrong.

You know it's true even though you don't like being reminded of it. The quickest way to get on someone's nerves is to make fun of their physique, but showing that their intuition is wrong is almost as good. In general, it provokes one of two defense mechanisms: people either say to themselves that they're losers, develop an inferiority complex, and stop thinking, or they say they're right after all and all the others are idiots (and stop thinking).

There is, however, a third way. When someone told Einstein or Descartes that their intuition was wrong, they didn't get upset. They didn't think they were idiots. They also didn't think that the others were the idiots. They reacted differently. How? It's one of the central themes that will reappear throughout this book.

In school, when they teach you to be wary of your intuition, they make two mistakes—two big mistakes that hold back your intellectual development.

The first is to exaggerate things. They get you all worked up over nothing. Sure, your intuition is wrong every now and then, but not always. Often it's right. And you can make it so that it's right more often. You can train it to see more clearly and distinctly. Starting from the same point as you, mathematicians construct a visionary intuition that is powerful and trustworthy. They get there using simple methods, like those taught in this book.

The second mistake schools make is to talk at length about the limits of intuition without ever reminding you of its strengths. The message that sticks with you is that intuition is imperfect. And that's an important message. But schools forget to pass on an even more important message: **your intuition is your strongest intellectual resource.** In a sense, it's your *only* intellectual resource.

These aren't just empty words. I'm not trying to flatter you to get on your good side.

Behind all this is hidden a profound biological truth that we'll talk more about later. It's also a very practical truth that you've experienced a million times before. You know that learning things by heart, applying ready-made methods, or following reasoning line by line isn't really understanding. That's why you never have complete confidence in logical arguments and you're much more at ease with what you understand intuitively.

The Gift of Imagination

You've known for a long time that your intuition is powerful. You wouldn't dare say it out loud, but it's really your intuition that you secretly rely on.

What you may not know is that behind all the great scientific revolutions and all the most difficult mathematical theories there are always intuitions, and these intuitions are always as simple as your own.

What allowed Einstein to come up with the theory of relativity was a mental cartoon not much more complicated than the one that let you see a straight line can't intersect a circle at three points.

When Einstein said he believed in intuition, he wasn't referring to a special form of heaven-sent intuition radically different from our own. If he'd really thought that, he wouldn't have said, "I have no special talent."

It's a bit disconcerting, but you have to accept the facts. Einstein was talking of everyday intuition, the kind that we all have, that which is often seen as childish and that school teaches us to distrust. Einstein was simply speaking of our ability to imagine things. It's a gift that we're all endowed with. You might think it's no big deal, but it's really quite something, and no one gets anything more than that.

If, like Einstein, you'd learned to use your simple and childish imagination to become the greatest physicist of your time, you would have said, as he did, that the great scientific discoveries are simply a matter of curiosity (and people wouldn't have taken you seriously).

And even if you haven't invented the theory of relativity, you've already done astounding things. You've been able to picture a circle in your head. You've been able to move it around with your mind. You've been able to visually prove to yourself that a straight line can't intersect a circle at three points.

And all that you've done by closing your eyes and staying still. You've been able to do it, literally, by the power of thought.

To the extent of our knowledge, this biological prowess seems to be limited to humans. If hippos also know how to do it, they're hiding it well.

If you've been able to do these things, rest assured: you have the genetic potential and the intellectual faculties to become very good at math. From the biological perspective, that's all that's needed. The other ingredients aren't genetic, and they're also at your disposal. It's simply a matter of sincerity, patience, desire, and courage.

Creating Clear and Strong Images

The big ideas are always intuitive and always simple. They're even ridiculously simple. We only ever really understand things that are obvious. When it's not obvious, it's because we haven't really understood.

This is a universal law of human cognition. It states that our science was invented by humans and that humans are, at the deepest level, all made of the same stuff.

The great discoveries are made by people who are simply trying to understand. They just want to make things clear for themselves. When they don't understand, they don't pretend they do. They continue to search for the right path, the right mental images, the right way of seeing, until it becomes obvious to them.

The good news is that with this method they can only discover things that are obvious. And what was obvious to them could someday become obvious to you as well. This is an excellent reason to try not to let yourself be intimidated.

That goes for all intellectual matters, but even more so for math. Mathematical knowledge isn't based on experimental data. It doesn't require amassing encyclopedic knowledge. In fact, it is entirely based on explicit proofs, which means that every result can be broken down into a succession of obvious deductions.

The paradox is that in order to get to the point where something becomes obvious to you, you first have to construct mental representations that allow that to happen. Once constructed, these mental images allow you to see it immediately and without any effort. But it takes a lot of time and effort to construct them.

Without realizing it, you've already constructed a good enough mental image of what a circle is. To understand math, you simply have to reproduce what you managed to do with circles with other objects, construct other mental images, and then combine these mental images to create yet others.

No one is born with these images ready-made. No one is able to construct them instantly. The process of constructing them takes more time than you'd think. For everyone, it's a matter of uncertainty, trial and error, false leads, and starting over again. And it goes on for your entire life.

Whether you do math or not, your vision of the world and your mental images are constantly evolving.

The oral tradition of mathematicians begins here. It's not a question of miraculous recipes for becoming superhuman, but of simple principles that foster the construction of better mental images.

What's at stake here is your ability to reclaim control of the way you construct your own vision of the world.

You know that to stay healthy you need to exercise, eat a lot of fruits and vegetables, stay away from drugs, and get a good night's sleep. But can you name the few basic principles that will help you construct strong and clear mental images?

This subject has never seriously been tackled. When everyone was trying to make you believe that you have to think logically, no one helped you develop your intuition.

You've made do without any method and under the false belief that your intuition is sometimes right and sometimes wrong, but that in the end there's nothing you can do to make it better.

In this context, it's a miracle that you've managed to learn anything at all.

And yet, as we'll see in the next chapter, you've done quite well. You've already managed to develop a solid mathematical intuition. You may think you're terrible at math, yet you've perfectly assimilated mathematical ideas that, for 99 percent of human history, seemed reserved for geniuses.

4
Real Magic

Take a billion. Then take away one. How much is left?

You don't really need to think. You can see the answer in your head: 999,999,999. The answer is actually easier to picture than it is to pronounce.

It seems obvious, and yet it wasn't always like that. To someone living in ancient Rome, for example, it wouldn't have been obvious at all.

In classical Latin, the word *billion* didn't exist (neither did *million*). To communicate the idea, the easiest thing would have been to call it the product of "a thousand times a thousand times a thousand." A Roman during the time of Julius Caesar should have been able to understand that, even if it might have given them a bit of a headache. But if you had told them that you were capable of taking this number, subtracting one from it, and picture the answer *immediately* in your head, they wouldn't have been able to follow.

They would have taken you for some kind of math whiz.

You'd be hard pressed to write 999,999,999 in roman numerals. If roman numerals are the only numbering system you know, 999,999,999 is much more than a big number you don't run into every day. It's a number that you can't even "look" at. It's so terrifying that it makes your head spin. The idea that someone could instantly "see" it clearly and without any effort is absurd.

But there's nothing extreme about the ancient Romans. Their understanding of numbers was really quite advanced. The traditional way of counting among certain aboriginal Australian people is based on parts of the body. You count from 1 to 5 on the fingers, then move up the arm: 6 is the wrist, 7 the forearm, 8 the elbow, 9 the biceps. When you get to 10 (the shoulder), you keep going up the body—12 is the earlobe. Yet if each number needs a corresponding body part, how do you get to a billion?

In the Amazon, Yanomami languages have an even more restricted numeral system: there's a word for "one" and another for "two," but there's no word for "three," just a catchall word that basically means "a lot."

For someone who sees the world in this way, discovering that there's a clear distinction between 25 and 26 that can be perceived in a split second must come as something of a revelation, comparable to what math students experience when they learn that there are many different sizes of infinity that can be precisely described.

A Complete Sham?

An inhabitant of ancient Rome would be able to grasp immediately the difference between XXV and XXVI. But your agility with big numbers would lead them to believe that you're a math whiz. That idea makes you smile, because you know for certain that you're no math whiz.

But are you sure about that?

If you think a math whiz is some kind of mutant with supernatural powers, if you think that they have some kind of computer in their head that lets them do calculations super quickly using the same methods that you know, then you're wrong.

In the end, math whizzes are kind of like Santa Claus: they don't really exist. When you think you've seen Santa, it's never really Santa, just someone dressed up like him. When you think you've seen a magician, it's never really a magician, it's always an *illusionist,* someone who knows tricks that can create the illusion that they have magical powers.

And when you think you see a math whiz, it's never really a math whiz, it's always just someone who has a way of seeing numbers that turns calculations that you find complex and scary into something easy and even obvious.

The truth is that we're all basically bad at mental calculation, except when we have an intuitive way of radically simplifying the calculation and "seeing" the result.

The decimal system based on Hindu-Arabic numerals is a "trick" that lets us see certain results as obvious. The main difference between a math whiz and you is that their bag of tricks is bigger than yours and they're more used to playing with them.

Real Understanding

The decimal system of writing numbers seems so obvious to you that you can't even remember learning it. It's just like using a spoon. You use it without really thinking about it, like it's an extension of your own body. When you see 999,999,999, you think you're seeing the number directly, without realizing that you're seeing it with the help of a tool.

Decimal writing is a purely human invention. More than simply a system of writing, it's a door into a state of consciousness where whole numbers, however big they may be, become concrete and precise objects. At the same time, the infinitude of whole numbers becomes commonplace.

Something previously unimaginable suddenly becomes commonplace: this is exactly the type of effect mathematics produces in your brain. It's a marvelous sensation, a great delight.

When you were a child, you were proud to be able to count to

10, then 20, then 100. It gave you bragging rights at recess. In order to brag some more, you would have wanted to know the biggest number.

To tell the truth, your awareness of numbers wasn't that far off from those people who can count to 2 or 5 and are firmly convinced that the next number, the number *many*, is the biggest number.

One day, you realized that no number was the biggest. Even if you might have arrived at this conclusion some other way, decimal writing gave you a shortcut. You know that every number is followed by another. You know how to see the succession of numbers like a counter that turns, and you know that this counter can turn indefinitely. There's no limit, there's no special number after which the counter stops working.

Yet for 99 percent of human history, no one had been able to picture a number counter turning in their head.

The number counter turning in your head is the collective work of great mathematicians who, from prehistory until the Middle Ages, fashioned the image of numbers that we share today.

This image isn't natural. It wasn't inscribed in your body the day you were born. It's partially arbitrary: we might have chosen another system for writing numbers, and you would see them differently.

More than four thousand years ago the Babylonians invented a *sexagesimal* system: they wrote their numbers in base 60 rather than base 10. Babylonian mathematicians were the most advanced of their time. Your mental image of hours, minutes, and seconds remains profoundly influenced by their vision of numbers.

What is natural, however, is your capacity to assimilate abstract mathematics and to really understand them, to modify your brain so that this math really becomes part of you.

You believe you can see the number 999,999,999. What you're really doing is deciphering a complex and abstract mathematical notation. You decipher it instantly, fluently, without even realizing it. Whole numbers may not be your mother tongue, but you've become bilingual.

Successful math becomes so intuitive that it no longer looks like math. If the example seems stupid to you, it's precisely because you understand it at the deepest level.

Real Magic Doesn't Exist

At the start of their careers, young mathematicians often feel like imposters.

It's a feeling I know well, and in my case it seemed entirely justified. The results contained in my PhD thesis were so obvious that it was almost like a trick. My theorems were always simple, and their proofs never contained any real difficulties.

Everyone around me seemed to be better at math. They were working on profound stuff that was way out of my league. They were writing papers that were extremely difficult to read, with proofs that seemed incredibly complex and technical. If I managed to understand a few of them, it's only because they happened to be easier than usual.

I wanted to know how to do real math, difficult math. But all that I was able to learn was the easy math, the math for dummies.

It seems silly to say this, but it really took me years to realize it was only an optical illusion. The horizon was shifting with me. It was always staying at my level.

Real magic doesn't exist. When you learn a magic trick, it ceases being magical. That may be sad, but you'd better get used to it.

If you find that the math you do understand is too easy, it's not because it's easy, it's because you understand it.

5
Unseen Actions

A great mathematician is, for example, someone born into a culture where people only know how to count to 5, and one day realizes that you can go further than that.

No one invented the infinitude of numbers out of the blue. At first, mathematical ideas are shifting and uncertain. You have the feeling that you might be able to go to 6 or 7, but you aren't able to articulate it because there are no words for "six" or "seven." You have the impression of being able to go even further, but this impression is fleeting. You don't completely believe it, you tell yourself something can't be right.

This is what happens when you run up against the limits of language.

In order to express what you feel, you have to invent new words, or create a new usage for words that already exist. Fleeting impressions cease being fleeting only after you find a way to pin them down with words. It takes time to get there. Words don't come easily, and they don't come right away.

The initial phase of a discovery is a spiritual experience. You think outside of language. The world is illuminated. You have epiphanies. You see things that until then were hidden. Things so new they don't yet have a name.

You know what I'm talking about. You've already experienced

this marvelous feeling. Try to recall your first time. It was the day of your first great mathematical discovery.

When you were a baby, long before you could speak, you probably played with a shape-sorting toy.

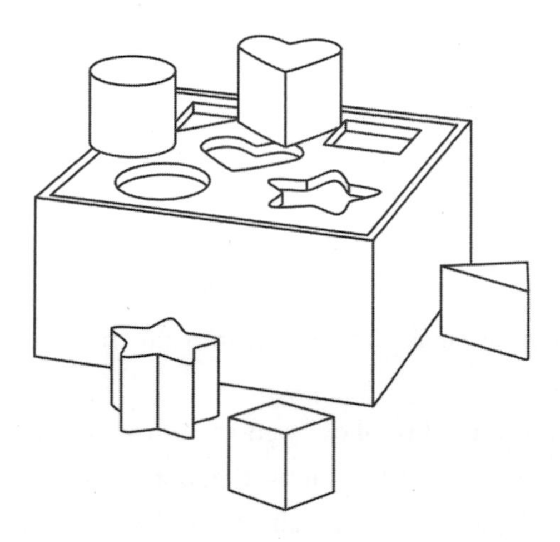

Your parents showed you how to do it. They took a block and put it in a hole. You wanted to copy them. You took a block and tried to put it in a hole. But it didn't go. You pushed with all your strength but it didn't budge.

That got you annoyed. Your parents told you that you couldn't force it in, that you had to look carefully at the shapes and match them: the round block in the round hole, the square block in the square hole. See, it's easy, right?

Except that you didn't understand what they were saying. You didn't really have any chance of understanding. The words *round* and *square* meant nothing to you. It wasn't just the vocabulary that you lacked, it was worse: you lacked the shapes themselves. You didn't know how to see them. Circles and squares were invisible to you.

All that you could see was that your parents were able to put the blocks in the holes and you weren't. And yet you were following their actions exactly. For them, these actions worked, for you, they didn't.

The scene was repeated dozens of times. Months went by; it was the greatest frustration of your young life. Your parents were magicians, not you. It was unfair and cruel. It made you mad.

But you didn't stop trying. You went back hundreds of times to this mystery that was so humiliating. Forget about the humiliation, you wanted to understand. You wanted in on the secret.

And then one bright day, you understood. You took a block in your hand and you noticed that this block had something particular about it, and that one of the holes had the same particular thing in common with the block. And it was into this hole that you had to place the block.

This realization didn't take any effort. You were just in the middle of going through the usual motions, the motions that even yesterday didn't seem to work. And then all of a sudden the answer seemed obvious. It was like your eyes were suddenly opened.

This was the period of your life when you invented the idea of shapes. It wasn't only about that block and that hole. It was about all blocks and all holes. Each block had its corresponding hole, and they shared this immaterial thing that didn't have a name. It worked every time. It was the secret of the magic trick.

You invented the idea of shapes by yourself and for yourself. It wasn't a preexisting knowledge, outside of you, that language brought to you. You learned all by yourself to see the shapes because, before you could see them, no one could explain to you what they were. You later learned the words for the shapes, but only after your perception of them.

From this point forward, you couldn't stop yourself from seeing them. It became ridiculously easy. Circles and squares, triangles and stars, hearts. It's actually a bit too easy for you. You've become incapable of imagining what it was like *not* to see them.

A Love Story

Let's set the record straight.

When you'd understood the secret of fitting the blocks into the holes, you were happy. You were immensely proud of yourself. You had a big smile.

Your parents were also proud and happy for you. They'd given you the toy, after all, to make you happy.

Maybe your parents didn't know it, but they really loved math. In giving you this toy, they wanted to pass along their love of math. And they succeeded. If you have kids, you'll also want to give them this toy.

Before school came along and got all caught up with it, before our inhibitions and our fear of being judged came along, we all have experienced great joy in math. Between humans and mathematics, it's been a long and profound love story.

Your beginnings were promising. Your discovery of shapes was really a great mathematical discovery. I'm being serious. That's not a metaphor.

Okay, it was a great discovery that was entirely pointless from a scientific perspective. You'd just rediscovered something everyone else was already aware of. But from the point of view of your personal knowledge, it was spectacular.

What you felt that day was exactly what mathematicians feel when they make a discovery. A mathematical discovery is just as simple, profound, and obvious.

Before Descartes, no one knew that you could describe geometric figures using equations. In *Geometry,* his 1637 treatise, an appendix to

Discourse on Method, he established a bridge between algebra and geometry, two branches of mathematics that had previously been thought of as entirely separate. These discoveries were the origin of the modern idea of *cartesian coordinates,* something that's since become obvious for any schoolkid: you can identify a point in a plane by providing its *x* and *y* coordinates. It's hard to imagine that before Descartes no one had "seen" cartesian coordinates. It's almost absurd, like imagining people couldn't see circles and squares.

Understanding a mathematical notion is learning to see things that you could not see before. It's learning to find them obvious. It's raising your state of consciousness.

When you look at the world, you can't help but recognize shapes, size, textures, colors. But there are many other things you might see. There are other structures, other types of shapes, other types of relations between objects. Even if you might not be able to see them now, these shapes and structures could eventually become obvious to you.

They're not that far away.

They're not that difficult to see.

They are *literally* right before your eyes.

Billie and Her Friends

Fitting the right blocks into the right holes isn't any harder than eating with a spoon. But *learning* how to fit the right blocks into the right holes is a whole lot harder than learning how to eat with a spoon.

In the case of the spoon, you can learn by imitation. With the game of blocks and holes, you tried to learn by imitation, but that didn't work. You were missing the critical step. Recognizing the shape of the block and identifying the right hole were unseen actions that your parents were mentally performing and that you weren't in a position to directly imitate.

You have to keep in mind that most of the things we learn, we learn through imitation. The instinct for imitation is universal. We share it with all other mammals, and not only them.

My favorite story of learning through imitation is that of Billie and her friends. Billie was a female dolphin who lived in the Port River, Adelaide, Australia. When she was young, Billie got lost. Isolated from her group, trapped in a lock and worn out, she was rescued and placed in a dolphinarium until she recovered her health.

There were captive dolphins in the dolphinarium that humans had trained to perform acrobatic tricks. Billie saw them and spontaneously began to imitate them.

Her favorite trick was the tail walk. That's when a dolphin swims fast underwater, on its back, then suddenly leaps straight out of the water. With the leap, the dolphin looks like it's walking backwards on its tail—thus the name tail walking. It's a very difficult maneuver,

physically intense, with no other function than pleasing humans or showing off in front of the other dolphins.

Three weeks later, when Billie was released back into her native habitat, she continued to tail walk. This trick had never been observed before in a wild dolphin. But the most interesting thing was what happened next: the other females in Billie's group started to do the same thing. Tail walking became all the rage among the dolphins of Adelaide.

In this way, we're exactly like the dolphins: not only are we able to learn by watching others, but what is more, we have an urge to imitate them. Our instinct pushes us to copy one another.

It's through imitation that we learn how to tie our shoes, use a toaster, ride a bike. We may not get it right on the first try, but watching others do it gives us an idea how. We know more or less what a shoelace or toaster or bike is for, and we know more or less how to use them.

But math, because it relies on unseen actions, can't be learned through imitation.

The Fosbury Flop

To make a mathematical discovery, you have to start by inventing for yourself new mental actions, creating new images in your head, without knowing in advance how to do it or whether it will work.

Inventing a truly new action is so rare in life that it's difficult to find well-documented historical examples. Even Michael Jackson didn't invent the moonwalk. He learned it by imitation. The origin of the dance moves goes back at least to the 1930s, and to this day its inventor remains an anonymous genius.

Dick Fosbury was someone who invented a new action: the high-jump technique that bears his name.

Before Fosbury, the two main techniques were the scissor jump (on the back, legs first) and the straddle jump (on the stomach, shoulders first).

The Fosbury flop, on the back and shoulders first, might seem counterintuitive. Taken out of context, with no cushion, it seems suicidal. Our body doesn't want to do that. Jumping backwards headfirst requires getting over our instinct that tells us that, clearly, this move is too dangerous to be attempted.

Fosbury didn't copy this movement from anyone else. He began to imagine it in 1963, when he was sixteen years old, and spent years perfecting it.

Fosbury would have been happy to copy someone else. He wasn't vain. He wasn't looking to be original or creative. He knew that imitation was the best way to learn, and naturally first tried to high jump like everyone else.

His starting point was in high school, where he was the worst member of the team. Because he wasn't successful with the official techniques, he began experimenting, looking for a more intelligent and efficient way of jumping: "It was not that I was trying to win, but I was trying to not lose."

The strength of his technique was that it allowed him to cross over the bar by rolling around it while his center of gravity remained under it: each part of the body successively passed over the bar but on average the body stayed below it. With the same thrust you can jump over a much higher bar.

Fosbury understood these scientific aspects. At university, he majored in civil engineering. But he discovered his method through introspection rather than calculation. He paid close attention to what his body was telling him, concentrating on the movements that would allow him to more easily cross the bar. Fosbury's approach was both deliberate and meditative.

One day, by modifying his run and the position of his body, he broke his personal best by six inches. It was his first real success in a school competition. It was then that he knew he was on to something. But his coaches didn't believe him. For years, they continued to try to convince him to jump the "right" way. Fosbury himself didn't really have any counterarguments for them. He just said that maybe his technique wasn't right, but it was right *for him.*

Thanks to his technique, Fosbury won the gold medal at the Mexico Olympics in 1968. He was twenty-one years old. His first interviews show that he himself didn't entirely understand the depth of his achievement: "I think quite a few kids will begin trying it my way now. I don't guarantee results, and I don't recommend my style to anyone."

But everyone copied him. From the next Olympics, in 1972, and onward, his technique became the norm. For more than forty years, every time a new record is set in the high jump, it's thanks to Fosbury's flop.

Blindly Copying

The most striking elements of Fosbury's approach will reappear throughout this book, as we tell the story of how mathematicians actually work.

A discovery always begins with the simple and innocent desire to understand. You invent new actions not because you want to do something new and original, but because you can't get where you want to be with the existing techniques. Without any reference point, without someone to guide you, you have to listen to what your body is telling you. You have to get used to feeling your body in a new way. Finding the solution means thinking what had been unthinkable. It's like augmenting the cognitive capacity of human beings.

A particularity about mathematics is that understanding a discovery is almost as challenging as making the discovery itself. In order to reproduce unseen actions, you can't avoid introspection. You have to listen to yourself, and reinvent the actions *within yourself* and *for yourself.*

To illustrate this, imagine an invisible version of the high jump that's done without an audience or cameras, in an empty room, with the competition judged by electronic equipment that verifies the bar was crossed without recording the jumper's technique.

How would Fosbury have been able to tell his story?

Everyone would have been convinced that he was genetically programmed to jump higher than anyone else. No one would have believed him if he'd said, "I have no special talent. I am only passionately curious," or "It was not that I was trying to win, but I was trying to not lose."

He might have written a book describing his technique and how it felt to him. But how to find the right words?

The first time one of his jumps was filmed, Fosbury said he was surprised himself, finding it hard to believe that what he saw on the screen was physically possible and really corresponded to what he'd done. For someone who's never seen it done, learning to jump like Fosbury is almost as hard as inventing it yourself. Even with detailed written instructions, it's not easy. "Jump up in the air on your back, headfirst." Seriously? And why all these preliminary pages about the trajectory of the run-up and the twisting of the body axis at the approach to the bar? Why all this technical language? Is it really necessary?

To truly learn a movement, you have to understand it beyond words. You have to feel it within your own body, and find it natural and intuitive.

Unseen Actions

Math is mysterious and difficult because you can't see how others are doing it. You can see what they're writing on the blackboard or on a sheet of paper, but you can't see the prior actions they performed in their heads that enabled them to think and write those things.

Math itself is simple but the mental actions that allow us to make sense of it are subtle and counterintuitive. These actions are invisible. We can't simply imitate what others do. We lack the adequate words to explain how to do it, and in any case words will always miss the main point: what we really feel inside ourselves.

All students must re-create the actions for themselves, blindly.

It's all too easy to make fun of math teachers, but try putting yourself in their place. How would you explain to someone how to tie their shoes if that person had never even seen shoes and your only means of communication was by phone? Take a few seconds to imagine the scene and you'll see how hard it would be. The very idea is so difficult it makes your mind reel.

This is the practical reality of teaching math, and we're all in the same boat. Professional mathematicians have this in common with people who are bad at math: they both know what it feels like to be totally at sea.

This feeling is part of their everyday life. Mathematicians at research conferences know that things will probably fall apart in the first five minutes. They know that it won't do any good to continue on, that it will just be sad and humiliating, because the words will simply have no meaning.

But they know that getting lost is a normal stage in the understanding process. They won't get upset. They won't pretend to understand what they can't understand. They won't even try to take notes. They'll simply stop listening.

If they really want to understand, they'll find another way.

6
Refusing to Read

I'm not a collector. I don't derive pleasure from accumulating stuff. That also goes for books, and a number of times in my life I've gotten rid of a big part of my library, giving away or selling most of my books. I kept only those I was especially attached to.

My math library is rather limited: fewer than one hundred books. Not many people have a hundred math books, but some mathematicians have a lot more. I've accumulated these books throughout my studies and career. Some of them I was given because I knew the author. Fewer than a hundred, after all these years, isn't all that much.

Most of the books I've needed, I've either borrowed or read the electronic version. I've bought only those that I've really liked, that I've really wanted to own, or that I found especially beautiful.

One of my favorite books, one of the few that it would break my heart to give away, is *Categories for the Working Mathematician* by Saunders Mac Lane.

Every time I come across it, I smile to myself. This book, first published in 1971, remains a reference in category theory, the revolutionary way of seeing and thinking about mathematical structures that Mac Lane and Samuel Eilenberg invented in the 1940s.

I bought it twenty years ago, just after I'd finished my PhD, with my first paycheck as an assistant professor at Yale University. There are few books that have touched me so deeply. I find it splendid, luminous, inspiring, and remarkably well written.

And I've never read it.

Raphael

When I started working on my PhD, even though I was officially one of those who is *very good at math,* even though my job already was to produce new math, I felt overwhelmed by the existing knowledge.

Every time I opened a research article, I got stuck on the first few lines. I was missing the basics. I looked for them in the references cited in the first article, but these references weren't any easier to read. So I searched in the references to the references.

The references, the references to the references, the references to the references to the references: it never ends. Even going back to the math from the 1950s, I found out that it was already incomprehensible to me.

Here I was, decades later, buried under thousands of books and tens of thousands of articles, none of which I understood. How could I hope to invent something original?

One day, I heard about a recent book on a subject that was useful—but not central—to my research. Everyone was saying that this book was very clear and well written. Which made me want to read it.

After a week, I hadn't even made it to the third page. Demoralized, I went to ask the help of my friend Raphael Rouquier, a young math prodigy I shared an office with.

His reaction remains engraved in my memory: "Come on, David! Didn't anyone ever tell you that you should never read math books? Didn't they tell you they're impossible to read?"

Daring Not to Read

No. No one ever had the guts to tell me so clearly.

Raphael was exaggerating: it is possible to read math books. But it's not natural and it requires a tremendous effort, even if you're very good at math. Reading a math book (and not just a book *about* math, like the one you're holding) is almost as difficult as writing one.

There's a good reason for that. When you open a math book, you're opening a book where the most important words have a meaning that you can't yet understand. This meaning might be specific to this very book. To be in a position to read it, you first need to find a way to make sense of these words. This requires constructing for yourself the right mental images for each word and each group of words, which comes at a heavy price. The effort will be almost as intense as that which allowed the author to write the book, and it will lead you to understand the matter almost as well as the author does.

If you really want to, if you have enough time for it and you've chosen the right book, it's well worth the effort. Get ready for a few months of hard work. This initiation rite will transform you. In my lifetime I've really succeeded in reading only three or four math books. I don't regret the time and effort. It gave me unexpected powers, as if I'd drunk a magic potion. This power remains with me today. But the potion was hard to swallow.

Even if Raphael exaggerated a bit, he was essentially right. Math books aren't made to be read.

Raphael was the first person I met who had never been afraid of math. It was no problem for him to take a five-hundred-page book on a subject he knew nothing about and open it right in the middle.

Even his way of holding books was different from mine. He didn't put his hands in the same place. Raphael balanced books on his forearm and held the top of the binding with the fingers of one hand, which left the other hand free to turn the pages very quickly. His technique was simple enough. He never started at the beginning, but wherever he felt like. It wasn't a reading technique as much as a non-reading technique.

When you pick up a math book, you always have an idea in mind. Maybe you want to understand an idea you've run into somewhere, to know if a certain statement is correct or not, or to get an idea of how to prove it. What really interests you might be Definition 7.4 on page 138, Theorem 11.5 on page 227, or maybe just a particular passage in its proof.

What Raphael taught me to do was to go directly to page 138 or 227 and find the four or five lines that at the moment interested me the most, *without having the least scruple* for skipping the mountain of preliminary material these few lines supposedly depend upon.

That's what's the most troubling. A math book is supposed to be organized logically, and to understand page 138 or 227, you *theoretically* need to have understood all that preceded it. Linear reading should therefore be the only possible way of reading. But *in practice* it's next to impossible.

In the four or five lines that interest you, there might be a few words that you don't understand. If that stops you from understanding the rest, you'll want to go back to the definitions. That's okay. Or you'll manage to muddle through anyway. That's okay too.

In fact, you should do whatever you feel like doing. You can leaf through the book for ten seconds, one hour, or three months—whatever. The underlying principle is never to force yourself to follow the pages in order, but to follow your own desire and curiosity.

The book should be at our service, rather than the reverse. It will never work if we try to read a math book like a "normal" book, if we let the book dictate the pace, if we wait for it to take us by the hand

and tell us a story. We're not there to listen passively. We don't have the patience—and, frankly, we're just not interested.

We're there because we have specific questions, because there are specific things we don't yet understand and that we want to understand. At any rate the book should never dictate the agenda. We're the ones asking the questions.

Let's face the facts. The four or five lines that interest us will be hard to understand, especially if they're in the middle of the book. It might take us hours. But every single page of a math book is equally hard. The so-called "easy" (but boring) preliminary pages are no less hard to understand.

In the end, the page that interests us the most may end up being the least difficult *for us.* First of all because we're interested in it: interesting things are a whole lot easier. And also because it's necessarily tied to something we already understand—otherwise it wouldn't interest us.

Following your desire is the only way of giving the book a real chance. If you start at the beginning, you run the risk of getting discouraged by page 2.

Bill Thurston

It's not only math books. There are other books that no one ever reads. Have you ever read the user manual for your toaster?

Probably not. You've probably glanced at it when you unpacked your toaster, but most likely you never opened it. Except, of course, if you had a problem with the toaster, in which case you skipped the beginning and went straight to the page you needed at that precise moment.

Comparing math books to a toaster manual may seem like a joke, but it's really a profound idea. And we owe it to Bill Thurston.

Bill Thurston, who was born in 1946 and died in 2012, is one of the most fascinating mathematicians of the recent era. His work in geometry, of exceptional depth and originality, constitutes a major step toward the proof of the famous Poincaré conjecture, achieved by Grigori Perelman in 2003. This work earned him the Fields Medal in 1982. Along with the Abel Prize, the Fields Medal is the most prestigious award in mathematics.

Thurston is among the greatest geometers of the twentieth century, but that's not all. It's not easy to sum up in a few words his mind-opening thought, his unique brilliance, his unlimited curiosity. To my knowledge, no other top-flight mathematician has collaborated with Japanese fashion designer Issey Miyake on the creation of an haute couture collection.

In 1994 Thurston published a twenty-page piece in the *Bulletin of the American Mathematical Society* that described his motivation as a mathematician and the mental processes at play in his work. Thurston notably said how, when he reads a research article in a field he's familiar with, he doesn't really read it. He prefers to concentrate on "the thoughts between the lines." Once he has a clear idea, the formalism and all the technical details suddenly seem useless and superfluous: "When the idea is clear, the formal setup is usually unnecessary and redundant—I often feel that I could write it out myself more easily than figuring out what the authors actually wrote."

"It's like a new toaster that comes with a 16-page manual," he continued. "If you already understand toasters and if the toaster looks like previous toasters you've encountered, you might just plug it in and see if it works, rather than first reading all the details in the manual."

The metaphor merits being stretched a bit. Thurston said that the manual is useless if you already know what a toaster is. But what if you've never seen a toaster before? Would the manual really be of any use?

You'd certainly be glad to learn that you shouldn't put your fingers inside it, and that you should never use it in the shower. But the blaring warnings that you're urged to READ CAREFULLY BEFORE USING aren't going to help you solve the great mystery of toasters: what are they used for?

I couldn't find my toaster manual, but I did find the one for my vacuum cleaner. It's sixty-four pages long and never says what you use a vacuum cleaner for. In other words, if we were to stick strictly to the official literature, vacuum cleaners would remain inexplicable. Nobody would know what to do with them, except the few folks who'd have received the "gift of vacuuming" (*that is,* they'd have accidently come up with a proper way to use them).

The hidden sense of vacuum cleaners isn't found in the manual. It's a secret we pass on by word of mouth.

What goes for vacuum cleaners also goes for mathematical theories. But this fundamental law of learning remains neglected and little known.

A Language That's Not Human

Math books aren't written in the language of humans. That's what makes them so hard to read.

The official language of mathematics doesn't work the same way as the language we speak every day, and no human could ever be perfectly bilingual. This artificial language is a purely human invention—undoubtedly one of the greatest inventions in our long history—conceived of to compensate for the weaknesses in the language we speak.

Its main particularity is to replace our usual way of defining words with a radically different approach.

In daily life we never correctly define the words we use. We learn them through examples. The best way to explain what a banana is, is to show one. This approach works well enough, but sometimes runs into issues, the most concrete of which is: when a thing exists only in your head, how do you point your finger at it?

At a profound level, math is the only successful attempt by humanity to speak with precision about things that we can't point to with our fingers. This is one of the central themes of this book and we'll come back to it a number of times.

In a math book, the most important passages aren't the theorems or the proofs: they are the definitions. Mathematical language works like building blocks where words are really *defined*: that is to say, built from other words that have themselves been previously defined. When you can't point your finger at something, that's a good way of going about it.

With this approach, the meaning of words is reduced entirely to the letter of their definition. Words are nothing but abstract shells that mean absolutely nothing outside the defined meaning: if having a trunk is part of the definition of an elephant, then an elephant whose trunk has been cut off immediately ceases to be an elephant.

This approach is called *logical formalism.* It's so anal retentive that it can be grotesque. We have no desire to think like that, and, furthermore, we're not really capable of it. Only robots and computers are crazy enough to do that.

It's the price you pay for speaking correctly about invisible things. Even if logical formalism is fundamentally foreign, we can learn to interact with it, in the same way that we can learn to interact with robots and computers. They may annoy us, we may find them ridiculous, but we end up getting used to their psychology and, at the end of the day, we're glad to have them work for us.

Learning to See

Learning math is learning how to use words that are "empty shells," defined by logical formalism, as if they were ordinary words. It's learning to give these words an intuitive and concrete meaning. It's *learning to see* the objects they point to as if they were right there in front of our eyes. This requires particular techniques that we'll talk about in the following chapters.

See isn't always the right word, since there are many material things you can't see. The taste of sugar, the touch of silk, a rhythm, a song, a familiar scent, the passage of time: these are things we feel without seeing them.

The ability to associate imaginary physical sensations with abstract concepts is called *synesthesia.* Some people see letters in colors. Others see the days of the week as if they were positioned in the space around them.

There's a widespread belief that synesthesia is rare and associated with certain mental conditions. In reality it's a universal phenomenon and a core building block of human cognition. Here's a little test to see if you're capable of synesthesia: looking at the word *chocolate,* are you able to sense a sound, a color, a taste? Looking at "999,999,999," do you get the feeling of something large?

What is rare, and what our culture doesn't push you to do, is to be aware of your capability for synesthesia and to try to develop it systematically. Secret math is a mental yoga whose goal is to retake control over our ability for synesthesia.

Nothing I'm saying should surprise you, since nothing about it is new to you. When you learned to "see" the number 999,999,999 rather than ink on a page, it was thanks to your command of this mental yoga.

What you could do in your childhood, you should still be able to do today.

Bill Thurston was a master of this art of seeing. In chapter 10, we'll talk about some of the things that he managed to see. It's so extraordinary that you'll find it hard to believe.

We have a lot to learn from him.

By Humans, for Humans

Now his comment on toasters should start to really make sense. When a human being looks at a math book, the trick isn't to read it from beginning to end, like a robot would. The trick is to get at "the thoughts between the lines"—that is, to give an intuitive sense to the words being used and the situations being described.

Math books aren't written by robots, for robots. They're written by humans, for humans. Without our ability to give them meaning, without "the thoughts between the lines," there wouldn't be any math books, exactly for the same reason that without music there wouldn't be any musical scores.

The best way of sharing this human understanding is direct communication between humans. This communication *about mathemat-*

ics is done *in human language.* As Thurston said, it's never as efficient as when there are only two people in the room: "One-on-one, people use wide channels of communication that go far beyond formal mathematical language. They use actions, they draw pictures and diagrams, they make sound effects and use body language."

When an important new theorem is proven, Thurston notes that the solution can often be explained in a few minutes during a private conversation between two specialists in the subject, whereas explaining the same result in front of an audience of specialists takes at least an hour. And to communicate the result in a written form requires a fifteen- or twenty-page paper that even a specialist may need hours, perhaps days, to understand.

Going from minutes to days is quite a leap. And what's worse, it's discouraging.

The Magic Touch

Mathematical understanding isn't some kind of magic power that can be passed along through touch, but it can seem like it. When you want to comprehend a mathematical concept, the quickest way is an open discussion with someone who really understands it.

Professional mathematicians are well aware of this. What worries them the most is their own difficulty in understanding mathematics. They have the same problem as everyone, but they know the solution.

When I was a PhD student, I didn't get ahead in my understanding by reading books. What I've been able to accomplish in my career I owe in large part to my conversations with Raphael. I was incredibly lucky to share an office with someone so good at math and so generous with his time.

Raphael's explanations were never rigorous. Sometimes they were plainly wrong. But they were always simple and human. They gave meaning. They made me want more. Raphael explained what a theorem *really meant.* He told me how an idea had been invented and how you need to understand it "morally."

"Morally" understanding something means being able to explain it to yourself intuitively and cite the reason why it's true: the *moral* of the story. If mathematics were simply a question of logic, no such thing would exist. There is no moral to draw from logical reasoning.

"Moral" explanations, waving your hands around, necessarily leave gray areas. They explain what toasters are used for and how to put bread into them, but they never detail their wiring diagrams. If that's really what you're interested in, feel free to go to Definition 7.4 on page 138.

The way of life and the social organization of the math research community reflect this need for direct conversations. Where astronomers have telescopes or nuclear physicists have particle accelerators, mathematicians also have their great scientific instrument: and this instrument is travel.

The trips mathematicians take allow the diffusion of new ideas with an efficiency otherwise impossible. We love long-term stays. We need time to talk, take coffee breaks, scribble on the blackboard, and pick up the discussion the next day with a question that came to us when we woke up. A Japanese mathematician, Kyoji Saito, wanted to understand the "thoughts between the lines" of one of my articles, so he often invited me to Kyoto. For my part, it allowed me to better understand the "thoughts between the lines" of his articles. These types of trips are part of a mathematician's life.

Clear and Intimidating

In the same math class where someone suffers in silence, there's probably another student who could explain things in simple language. Why does this conversation almost never take place?

The certainty of people "bad at math" that they're naturally inferior plays a role. They're too inhibited to ask the right questions, the really simple questions, the ones that seem stupid but that in fact are fundamental.

Teachers share in the blame. They sometimes foster the illusion that math can be limited to formal equations. Just as there's an incredible gap between those "good" and "bad" at math, there's an equal gap between "good" math teachers and "bad" math teachers.

Let me try to explain. In the world of mathematics, toasters arrive disassembled. We all have to put them together in our own heads. "Bad" teachers are the ones who recite the 198 steps to assemble the toaster as if that were the end of the story. "Good" teachers do their best to explain what a toaster is. They constantly look their students in the eyes, because it's in their eyes that they will know if they've understood.

Inflicting the 198 steps of putting together the toaster on someone who doesn't even know what bread is for is just plain mean. It's like raising children without telling them stories. You can't teach humans the same way you would teach robots.

I don't think that "bad" math teachers are deliberately sadistic. Maybe they don't put human understanding front and center in mathematics because they themselves haven't been taught the right way. Maybe they don't picture the toaster in their own head. Or maybe it's the opposite: when you see the toaster so well in your own head, sometimes you forget that not everyone else sees it the same way.

"One person's clear mental image is another person's intimidation," Thurston wrote.

It's hard to share mental images, as they are evanescent and profoundly subjective. Our common language is incapable of transcribing them with precision. It's precisely because our intuition is so secretive and so unstable that logical formalism was invented.

For Thurston, as for most creative mathematicians, mathematics is a sensual and carnal experience that is located upstream from language. Logical formalism is at the heart of the apparatus that makes this experience possible. Math books may be unreadable, but we nevertheless need them. They're a device that we rely on in our quest for the true math, the only one that matters: the secret math, the one that lies in our head.

Which brings us to a natural question. How do people find the courage and desire to write books that are unreadable, that readers couldn't care less about, and that are as dry as the user manual for a toaster? What's their motivation? What state of mind gives birth to mathematical creativity?

This is the subject of the next chapter.

7

The Child's Pose

"My dear Serre, Thanks for the various papers you've so generously sent me, as well as for your letter. Nothing new here. I've finished my ridiculous piece on homological algebra."

This is how Alexander Grothendieck begins a letter written to Jean-Pierre Serre on November 13, 1956. The casual tone is a bit surprising, especially when you know who Serre and Grothendieck are, and what the letter is about.

Jean-Pierre Serre is one of the greatest mathematicians of the twentieth century. A career isn't entirely measured by the awards you win, but when you've won them all, that's saying something. Serre won the Fields Medal in 1954 at the age of twenty-seven, and is still the youngest ever to do so. This prize is restricted to mathematicians under forty, and for a long time there wasn't anything equivalent to a lifetime achievement award in math. This is why the Abel Prize was created in 2003. The first year of the prize, the awards committee had a great responsibility: among all living mathematicians, they had to choose who was to receive the first award. They decided to give it to Serre.

As for Grothendieck, he is much more than a great mathematician. Well before his death in 2014, he'd already become something of a legend.

He's one of those rare mathematicians—one of the few throughout history—whose contributions aren't limited to profound results

or spectacular theories. Grothendieck invented a way of approaching math that was so rich and new it was as if he'd changed the very nature of mathematics.

This explains why he's often considered "the" greatest mathematician of the twentieth century, insofar as that means anything.

As for the "ridiculous piece on homological algebra," that was the article "Some Aspects of Homological Algebra" that appeared in 1957 in a Japanese scholarly journal, the *Tohoku Mathematical Journal.*

This article marked Grothendieck's entry in the field that would make him famous. Influenced by Serre, he was just getting started in algebraic geometry. The two young men began one of the most productive mathematical friendships in history. Of his first encounter with algebraic geometry, Grothendieck would later say that he had the impression of "suddenly finding himself in a sort of 'promised land' of luxuriant richness."

Grothendieck would devote fifteen years of his life to mapping out this "promised land." Writing was at the heart of his method. He would go as far as to say that "doing mathematics is above all *writing.*"

This passion for writing makes his letter to Serre even more enigmatic. The "ridiculous piece" is quite simply the first tale of Grothendieck's adventures in the promised land. He found the piece so ridiculous that finishing the task that had consumed him for a year was a nonevent. "Nothing new here" are the exact words he used to announce that he'd just finished writing a historic article.

Was he joking? Probably not. In a 2018 interview Serre recounted that one of Grothendieck's peculiarities was his total lack of humor: "I can't recall having heard him laugh. You could never joke with him, for example, about mathematics."

Hidden behind this apparent paradox lies a profound truth about the nature of mathematical work. Grothendieck's detachment and flippancy might seem incomprehensible, but when you know more about his intellectual approach, you'll find it perfectly coherent.

A Joke in Bad Taste

Everyone knows who Einstein is, but almost no one's heard of Grothendieck.

Comparing the two isn't absurd. Einstein revolutionized physicists' ideas about space. Grothendieck revolutionized mathematicians' ideas about space. He even went so far as to reinvent the concept of a point, and to approach the notion of truth from a geometric perspective.

Some mathematicians even think that the comparison with Einstein isn't fair to Grothendieck. They find Einstein's work beautiful, elegant, brilliant, admirable. They say that it's the work of a genius. As for Grothendieck's work, they find it extraordinary, staggering, sublime, terrifying. They say that it can't be the work of a human being. Grothendieck's ideas are not always easy to understand, but once you understand even a bit, it seems incredible that anyone could have come up with them.

Jean-Pierre Serre once said about Grothendieck's work that he personally would have been incapable of producing it, because "it demands enormous strength." When Serre evokes Grothendieck, he talks of "the power of his brain" and describes a supernatural force: "Physically and intellectually, it was the same. It was extraordinary. I've never known anyone with as much strength. I've met people with incredible intellectual abilities, but Grothendieck was beyond that. He was a beast."

Grothendieck himself wasn't of the same opinion. He didn't think that he was more gifted than anyone else. That wasn't the source of his uncommon creativity: "This power is in no way some extraordinary gift—like an uncommon cerebral strength, (shall we say). . . . Such gifts are certainly precious, worthy of the envy of people (like me) who haven't been blessed with them at birth, 'beyond measure.'"

Grothendieck had a different explanation: "The quality of the inventiveness and the imagination of a researcher comes from the quality of his attention, listening to the voice of things."

It almost sounds like the same words from Einstein that we started with in chapter 1: "I have no special talent. I am only passionately curious."

But Grothendieck went even further. He knew, however, that no one would believe him, since these kinds of declarations are never taken seriously: "When you dare say such things, you see on everyone's face, from the dullest who are sure they're dull, to the smartest who are certain they're smart and well above common mortals, the same smiles, part embarrassed, part knowing, as if someone had made a joke in bad taste."

Einstein had the reputation of being something of a joker. With Grothendieck, no need to worry: jokes were not allowed.

It's too bad we never could have had that conversation with Einstein, the one in which he would have spilled all the secrets of his creativity, the one in which he would have agreed to answer our questions and explained in detail *how he really did it.*

As for Grothendieck, he wrote a thousand-page book on the subject. He described in detail what went on in his head when he did mathematics. He acknowledged his total inability to read any math book, even the most simple, if he didn't manage to fabricate the right pictures in his head. He also acknowledged his inability to follow along during conferences because they always went too fast for him. He explained his way of getting by with the feeling of not knowing anything. And most of all he explained the exact place where he was finding pleasure in all that.

This extraordinary tale is called *Harvests and Sowings* (*Récoltes et semailles*). The manuscript remained unpublished for a long time, circulating clandestinely for over thirty-five years until someone finally dared to publish it. The first legal edition appeared in 2022, published by Éditions Gallimard in Paris. A forthcoming English-language edition is being prepared by MIT Press.

The Most Breathtaking Account

"What drives and dominates my work, its soul and reason for being, are the mental images formed during the course of the work to apprehend the reality of mathematical things. . . . All my life I've been unable to read a mathematical text, however trivial or simple it may be, unless I'm able to give this text a 'meaning' in terms of my experience of mathematical things, that is unless the text arouses in me mental images, intuitions that will give it life."

Many such illuminating excerpts can be isolated from *Harvests and Sowings*, one of the most fascinating books I've ever read. Yet I can't really recommend it, or at least not without a serious caveat: Grothendieck's monologue is a long and disconcerting rant, alternatively brilliant and confused, with prophetic overtones, interlocking metaphors and allegories, notes and digressions strewn about the bottom of the page, all with their own notes and digressions. He gets lost for hundreds of pages in personal grievances and unfounded recriminations that are, quite frankly, almost unreadable.

It's a text meant for the initiated, and even the initiated have a hard time making it through to the end. A frequent opinion, however, is that *Harvests and Sowings* is the most breathtaking account ever written about the mathematical experience.

Like many of my mathematician friends, I've found stunning passages of truth and clarity, moments when I stopped reading and said to myself: "He's right. That's it exactly. That's really it, the secret. That's how it really happens in our heads. It's precisely in making these simple mental actions, which seem so innocent but which no one had dreamed of doing, that you get really good at math. I've never read anything as important. I need to find a way to tell the world and explain what Grothendieck is trying to say."

I'm aware, however, that in the raw, Grothendieck's thought is far too enigmatic. In the end, it's a bit like the problem with Einstein. There's no longer the possibility of having a frank and direct conversation with him, to ask the simple and naïve questions.

Grothendieck went considerably further than Einstein and left us incredible details. But he did so as an isolated man, ahead of his time, aware that people were not yet ready to hear his message. To make sense of his account, we need to relate it to our common experience.

Before sharing with you what I've personally taken away from my reading, what has echoed my own experience, I should tell you a bit more about Grothendieck's life and his uncommon personality.

A Wild Child

Alexander Grothendieck was born in Berlin in 1928. His parents were militant anarchists who had to flee the Nazi regime. In 1933,

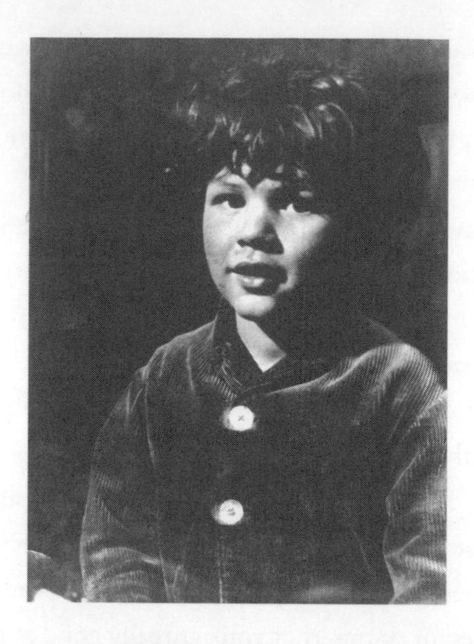

when he was five years old, his mother entrusted him to the care of the family of Wilhelm Heydorn, a Lutheran pastor in Hamburg.

Up until this time, Grothendieck seems to have received a rather unusual education, formed by the anarchist principles of his parents. His adoptive mother, Dagmar Heydorn, described him upon his arrival as a wild child, dirty and lacking any form of inhibition. When she entrusted the Heydorns with her son, Hanka Grothendieck did so on the condition that they never send him to school, and never cut his hair.

The Heydorns cut his hair and sent him to school. It was probably the only peaceful and "normal" period of his life. He retained throughout his life strong ties of affection with his adoptive family.

In April 1939, fearing for his safety (Grothendieck's father was Jewish), the Heydorns put him on a train to Paris, where he reunited with his exiled parents. His father was arrested shortly afterwards, and would die in Auschwitz in 1942. Beginning in 1940, Grothendieck and his mother lived in the South of France, in refugee camps.

Their life began to resemble clichés from Hollywood movies: the refugee mother and son in postwar France, scratching out a living doing housework or harvesting; the son's spectacular talent forming in isolation, without anyone taking notice.

In 1948, the atypical student was finally spotted by one of his professors at the University of Montpellier, who wrote him a letter of recommendation to Élie Cartan, a prominent mathematician with great influence. This is how at twenty years old Grothendieck made his way to Paris and was introduced to cutting-edge mathematical research by some of the most brilliant minds of his time.

Laurent Schwartz, who himself was about to receive the Fields Medal, had him read his latest article, which ended with a list of fourteen problems that he had been unable to solve. It's the kind of list an ambitious student could dig through for a good PhD subject: choose a problem, spend three years thinking about it, get your advisor to help you find an incomplete solution, and everyone's happy. Grothendieck went off to his room and came back a few months later. He'd solved all fourteen problems.

Until 1970, Grothendieck climbed ever higher on the ladder that separates an unknown refugee from the heights of global science. He became the greatest, the strongest, the best. He worked harder than anyone else. A whole research institute was created around him. He won the Fields Medal in 1966, but that is merely anecdotal in comparison to everything else he accomplished. Grothendieck and his students set out on the immense and visionary task of reconstructing algebraic geometry from the ground up. Their work still provides the basis for a large part of current mathematical research.

But in 1970, at the age of forty-two, Grothendieck made a sudden break from his scientific career. He resigned from the institute created around him and opened a new chapter in his life, dedicated to militantism and radical ecology.

It was in the mid-1980s, about fifteen years after this break, that he wrote *Harvests and Sowings.* His intention was to write a book for the general public because he believed he had an important message to deliver. In a letter he wrote in 2010, he acknowledged that he didn't entirely succeed: "This 'Reflection and Testimonial' on my life as a mathematician, unreadable as it is I admit, has much meaning for me, if not to anyone else!"

From 1991 until his death in 2014, Grothendieck retired from the world. He lived as a recluse in the small village of Lasserre in Southern France at the foot of the Pyrenees Mountains, where he practiced meditation and led an existence of extreme solitude and asceticism. He went as far as trying to live entirely on dandelion soup.

Grothendieck never stopped writing. He left behind him immense quantities of mathematical, philosophical, and mystical writings, among which, it seems, is a thirty-thousand-page meditation on "the problem of Evil."

The connection of the mathematical experience with madness is a subject we can't seem to ignore. We'll come back to it in chapter 17.

"The gift of solitude"

"Discovery is the privilege of the child. I'm talking about young children, children who aren't yet afraid to make mistakes, to look like fools, not to be serious, not to act like everyone else. They're also not afraid if the things they're looking at have the bad taste of being different from what was expected from them, what they were supposed to be."

This quotation from *Harvests and Sowings* sounds like something we've heard thousands of times before, but that is clearly not true. And even if it were true, what good does that do us? We'll never be young children again.

But it's obviously a metaphor. Grothendieck is alluding to the child who is present "within us" and with whom "we have lost contact." His book is actually addressed not to us but to the *lost child within us,* as he makes perfectly clear from the onset: "It's to the one within you who knows how to be alone, to the child, that I wish to speak, and to no one else."

Grothendieck explains his uncommon creativity by the proximity he maintains with his inner child: "In me, and for reasons I have not yet dreamed of exploring, a certain innocence has survived."

He describes this as a "gift of solitude," the capacity to find himself "alone and listening to things, intensely absorbed in a child's game."

"Seeking and finding, that is to say, questioning and listening, is the simplest, the most spontaneous thing in the world, that no one has sole rights to. It's a 'gift' that we all received in the cradle."

Whatever one thinks of Grothendieck, of his strangeness, his eccentricities, his bizarre obsessions, it's worth listening to him. Whether or not he's using the right words, he clearly knows what he's talking about.

Harvests and Sowings often reads like a yoga manual, and in a way that's exactly what it is. Behind the metaphors and personal anecdotes, the text describes a certain way of holding your body, a peculiar physical attitude, an unusual relationship to language and truth.

Grothendieck was a great yogi who invented his own meditation technique. It's centered on a radical form of curiosity and indifference to judgment, what we might call *the child's pose.*

All mathematicians develop techniques of this sort, but they're rarely aware of it, and rarely know how to explain it. Grothendieck hands us the user's manual.

This psychic posture is clearly at the heart of his work method. Roughly, here's what it consists of.

"I believe, more or less, in my assertions"

Opening a math book on a subject you know nothing about is a bit like finding yourself in the pilot's seat of a commercial jet or the command post of a nuclear generator. There are a lot of buttons and screens, but you have no clue how they work and an intense desire not to make a mistake. You would love to know how it all works, but you don't. The normal reaction is to stay seated and not touch anything. Before making any move you need to study and think about it.

But if you put any two-year-old in the pilot's seat, they'll act differently. They'll push all the buttons, starting with ones that are red or blinking.

Grothendieck's recommendation is to act like the two-year-old. When he wants to understand something, he goes straight at it, without hesitations, as a child would. He doesn't wait to understand before launching into it. He acts without thinking, a bit haphazardly:

> When I'm curious about a thing, mathematical or otherwise, I **interrogate** it. I interrogate it, without worrying about whether my question is or will seem to be stupid, certainly without it being well thought out. Often the question takes the form of an assertion—an assertion which, in truth, is an exploratory probe. I believe, more or less, in my assertions. . . . Often, especially at the outset of my research, the assertion is completely false—still, it was necessary to make it to convince myself.

We need to clarify what he means by *interrogating things, asking questions,* and *probing.* Throughout his book, Grothendieck describes mathematical work as a succession of concrete physical activities. But what does *interrogating things* mean exactly? If I want to interrogate things, how do I go about it? Looked at closely, this isn't clear at all. The same can be said about this other favorite phrase of Grothendieck: *listening to the voice of things.* What does that even mean?

As long as we're here, what is a *mathematical thing?* Where can we find these things, and how do we initiate contact with them?

Grothendieck never bothers to explain precisely, undoubtedly because he is so used to talking with these things that he's forgotten that he himself had to learn how to do it.

Mathematical things are the things that nonmathematicians call mathematical *concepts* or mathematical *abstractions.* They may consist of numbers, sets, spaces, different kinds of geometric shapes, or other types of abstract structures. Mathematicians prefer to call them mathematical *objects,* because imagining these things as material objects that one can touch makes it easier to understand them.

Interrogating things, listening to the voice of things, means trying to imagine them, examining the mental images that form within you, seeking to solidify these images and make them clearer, working at unveiling more and more details, as when you try to recall a dream.

The Pleasure of Being Proved Wrong

This approach needs to be translated in concrete terms. The language of *Harvests and Sowings* is so imagistic that you might think keeping it vague was a deliberate choice.

This impression is false. Grothendieck endeavored to be precise. His enigmatic vocabulary was meant to solve a practical problem: he's describing actions that we perform in our heads and mental images that we manipulate, but our language is missing the right words. There is no specific vocabulary to talk plainly of these actions and images. No one has taken the time to even tell us we have the right to talk about them.

The child's pose isn't an allegory. It's a very precise mental attitude.

The basic principle is simple yet revolutionary. It's the kind of idea that almost no one thinks of because it's too simple and it goes against our instincts. The kind of idea, precisely, that has the potential to change everything, at all levels of math learning, including the absolute beginners and the self-professed lousy at math.

When you come across a new mathematical concept, it's hard to imagine it. It is presented to you by means of an abstract definition, a string of words on a page or words spoken by a professor. This string of words doesn't make any sense to you. It has no intuitive meaning.

Students generally don't feel like they have the right to imagine mathematical objects that they don't yet understand. They feel they need to know more before daring to picture them. In the meantime, they're content to try to decipher word by word, symbol by symbol. They may not understand what they're reading, it might give them a headache, but they tell themselves that if they keep on trying they'll get to the point where they feel confident enough to finally imagine what's behind the words. But this approach almost never works.

Grothendieck did it differently. He knew that it was worthless to gather information about things that you can't yet see. Instead, he allowed himself to imagine the things right away, without waiting, even when he was well aware that it might not work and his mental images would likely be terribly wrong.

He wasn't afraid of failure. He was even certain that he would be wrong, and that's exactly what he was looking for.

Grothendieck actively sought out the error as a young child actively seeks mischief. In his exploration of the world of mathematics, each time he found something bizarre or intriguing, unclear or unsatisfactory, incoherent or disagreeable, that's where he began digging.

When something was off in his vision of the world, it made him feel uneasy. He dug around to find the source of this unease, since that was the only way to relieve it. Finding mistakes gave him pleasure, relief. "Finding mistakes is a crucial moment, above all a creative moment, in all work of discovery, whether it's in mathematics or within oneself. It's a moment when our knowledge of the thing being examined is suddenly renewed."

What Grothendieck wrote about error is of universal significance, well beyond the field of science. It makes you want to engrave his words on school façades:

> Fear of mistakes and fear of the truth is one and the same thing. The person who fears being wrong is powerless to discover anything new. It's when we fear making a mistake that the error which is inside of us becomes immovable as a rock.

Not many people realize that the main obstacles in mathematics are psychological, not only at the beginning but all throughout, up to the highest levels. As we leave childhood behind, we learn to fear looking stupid. We learn to be ashamed of our mistakes. We learn to hide, even to ourselves, the fact that we know almost nothing. To get ahead in math, we need to deactivate this reflex for dissimulation. And it's not easy.

At the age when we were still free to ask stupid questions, even to ask the same stupid questions hundreds of times in a row, no one hated math. The great mathematicians invent and put in place special techniques to recover this lost childhood innocence. They all say it's indispensable. We'll come back to this in chapter 13.

The Driving Force of Learning

When Grothendieck talks of "the error which is inside of us," that has nothing to do with logic. It's not an error of computation or reasoning. The error Grothendieck is talking of is one of intuition, an error of vision: the image that we have of things isn't correct.

As we'll see throughout this book, mathematical understanding is achieved by gradually modifying the way we represent things to ourselves, and making them clearer, more precise, closer to reality.

You sometimes hear people say that the two hemispheres of our brains function differently. The left side of the brain specializes in logical reasoning and calculation, while the right side specializes in associative reasoning and intuition.

This nonsensical interpretation of our anatomy dates from the 1960s and has long since been discredited. In reality the two sides of our brains closely resemble one another and, at a profound level, both function associatively and intuitively. The organ that allows you to see the world in a logical manner doesn't exist. If you're counting on that to become good at math, you'll have a long wait.

Our prodigious faculty for learning and invention has its origin in our unconscious ability to constantly reconfigure the fabric of associations of images and sensations that, literally and figuratively, comprise the real structure of our thought.

All our great learning achievements rely on this mental plasticity. Error plays a fundamental role, as it is the driving force of plasticity. Learning to see, to walk, use a spoon, tie your shoelaces, talk, read and write, is always about reconfiguring your brain. And it's never done in one shot. Children don't learn how to walk until they've tried and failed. They need to fall in order to learn how to stand up. It's the accumulation of errors that allow them to develop their intuitive sense of balance.

As for every motor learning, understanding a new mathematical concept proceeds by a reconfiguring of intuition, and that requires a "feeling-out" phase. Once transposed to the context of walking, Grothendieck's comments on the role of error become even more illuminating:

Fear of falling and fear of walking are one and the same thing. The person who fears falling on their face is powerless to learn how to walk.

It's only when we stay on our ass that our initial clumsiness turns into physical disability.

The Role of Logic

In the world of mental images, the laws of physics don't apply. You can imagine anything, even inconsistent things, without falling on your face. The error that is inside of us can become immovable as a rock without our even being aware of it.

It's precisely here that the mathematical approach diverges from our usual way of using our intuition. Mathematicians have invented a method that lets them discover the errors inside themselves. This method relies on writing—more precisely, on writing in the official language of mathematics, constructed on logical formalism.

Logic doesn't help you think. It helps you find out where you're thinking wrong.

When Grothendieck sends out "probes" to interrogate objects he wants to understand, he gets his answer by writing:

> Often, you only have to write it down for you to see it's incorrect, whereas before writing there was a vagueness, a bad feeling, instead of this evidence. That now allows you to start over without this lack of knowledge, with a question-assertion perhaps a little less off the mark.

As opposed to biologists, who write their articles only *after* having done their experiments, mathematicians write *during* their research work, because writing is itself part of the research. Here's what Grothendieck says:

> The role of writing is not to record the results of research, but is the process itself of research.

> I have always made every effort to describe in the most meticulous way possible, using mathematical language, these images and the understanding they bring. It is in this continuous effort to articulate the inarticulable, to define what is as yet unclear, that the particular dynamic of mathematical work (and perhaps as well all creative intellectual work) is perhaps found.

Mathematical writing is the work of transcribing a living (but confused, unstable, nonverbal) intuition into a precise and stable (but as dead as a fossil) text.

Or, rather, it would be a simple job of transcription if the intuition was from the outset precise and correct. But intuition is rarely precise and correct from the outset. At first it's vague and wrong, and it always remains a bit so. Through the work of writing, intuition becomes less and less vague and less and less wrong. This process is slow and gradual.

Mathematical creation is a constant back-and-forth between imagination (the art of picturing what you read) and verbalization (the art of putting words to what you see). This simultaneously transforms our intuition and our language. We learn to see and, at the same time, we learn to talk. We learn to picture new things and we invent a language that allows us to name them. The whole process, according to Grothendieck, amounts to "gathering intangible mists from out of an apparent void."

The result of this work shows up in two different ways. The first incarnation is invisible: it's the modification of the understanding of the world and the state of consciousness of the person who produced the work. The second incarnation is the mathematical text.

Grothendieck knew that this second incarnation, the printed one, was the only one he could show. But it wasn't what drove him. For him, "it's not in this form that you find the soul of understanding mathematical things."

The effort of writing allowed Grothendieck to develop his own intuition. Once he had a clear idea, he could look at his own articles with detachment, as if they were user manuals for toasters.

The Diplodocus

In the next chapter, we'll see how the peculiarities of mathematical language make it an incredible tool of mental clarification.

But we'll end this chapter by going back to the mystery we began with: the casual tone of the letter that the young Grothendieck, at age twenty-eight, wrote to Serre on November 13, 1956, to announce that he'd just finished his "ridiculous piece."

In June 1955, seventeen months earlier, Grothendieck had written to Serre to share with him his first notes. The tone was enthusiastic, as Grothendieck was in the initial phase of discovery. He probed into things, made massive errors, and achieved rapid progress. At the time he still qualified some passages in his notes as "unlaid eggs" that were potentially "screwed up."

In the year that followed, Grothendieck brooded upon his "eggs." He watched them hatch and patiently fed the bizarre creature that emerged. As the manuscript grew and gained structure, Serre and Grothendieck spoke about it with a growing glibness, going so far as to give it a nickname: the "diplodocus."

The big ideas were in place. The pleasure of discovery, the pleasure of finally understanding, was already fading. Surprises were becoming rare. It was simply a matter of putting on the finishing touches, the technical details, of conforming to the bureaucratic requirements of the official language of mathematics.

In the final months of editing, the writing became an ordeal. Grothendieck began to worry whether anyone would want to publish his ridiculous piece. He chose the Japanese *Tohoku Mathematical Journal* because "it seems that really long articles don't bother them."

In his letter to Serre, Grothendieck goes as far as excusing himself. He may have created a monster, but he didn't have any choice: "It's the only way I have of understanding, through sheer persistence, how things work."

8

The Theory of Touch

Among the books one never reads, apart from math books and user manuals for toasters, one mustn't forget dictionaries.

When I was a child, I was fascinated by dictionaries. Their promise is to define every word using other words. But do they keep this promise? Can they really function as a gateway to language? When you want to learn words from the ground up, which page do you turn to?

If you don't know what a banana is, the dictionary will teach you it's "an elongated curved tropical fruit of a banana plant, which grows in bunches and has a creamy flesh and a smooth skin; in particular, the sweet, yellow fruit of the Cavendish banana cultivar." But what is a banana plant? It's "the tropical tree-like plant which bears clusters of bananas, a plant of the genus *Musa* (but sometimes also including plants from *Ensete*), which has large, elongated leaves."

That's not wrong, but it's pretty confusing. The definition is bizarrely tortured and complicated, and most of all, it's circular: the banana is the fruit of a banana plant, which have bananas for fruit. Why not cut to the chase and just say a banana is a banana, as this seems to be the main message?

You don't explain what a banana is to someone who doesn't know with a bunch of convoluted phrases. To show what we really think about bananas, the simplest and most honest definition is still the one we give to children: "Try it! It's good!"

Dictionaries are filled with circular definitions.

What is heat? "The condition of being hot." What is hot? "Having or giving off a high temperature." What's a temperature? "A measure of cold or heat, often measurable with a thermometer." What's a thermometer? "An apparatus used to measure temperature." What is truth? "Characteristic of what is true." True? "That which conforms to the truth."

From a logical standpoint, dictionaries are giant Ponzi schemes. If people truly relied on them to find out about bananas, the fraud would have been denounced long ago.

But that's not how we do it. Our approach isn't logical. We don't learn words through their definitions. We assimilate language bit by bit, by successive clarifications. Our brain has the ability to see things before knowing how to name them, to recognize words before understanding their meaning, and to gradually associate the words with what we see.

We start from zero, literally. We don't start from dictionaries. We start with life and the common experiences we share with others.

Starting from Zero

Mathematical definitions resemble definitions in dictionaries, with one slight difference: they actually define things.

As opposed to dictionaries, math books don't simply make connections between words that already exist. They don't limit themselves to things you can point to or that we have a shared experience with.

A mathematical definition is neither a commentary nor an explication: it is the exact assembly guide of a new mental image and the "birth certificate" of the new word chosen to designate it. (In practice, existing words are often reused, receiving new meanings that may have no direct relation to what these words mean in everyday life.)

In that sense, mathematical definitions have the power of creation: they bring things into existence. It may seem silly to speak so pompously, but that's what really happens.

When you see things that others don't yet perceive, sharing your vision requires finding a way to get others to re-create those things in their own heads. A mathematical definition serves this purpose. It provides detailed instructions allowing others, starting with things they are already able to see, to *mentally construct* those new things.

An Enormous Factor of Expansion

In theory, everyone should be able to read math books. Unlike dictionaries, they contain no circular definitions. No implicit knowledge is required, and whenever it is necessary readers are referred back to previous references where they can find the definition of words they're not yet familiar with. Since the instructions are clear and all the details are given, there shouldn't be any obstacles to understanding.

In practice, however, from the beginning lines of a math book we're faced with a huge problem: it's fantastically difficult to explain a mental image in words.

As Thurston remarked, "There is sometimes a huge expansion factor in translating from the encoding in my own thinking to something that can be conveyed to someone else."

The result is often unpalatable. When Thurston speaks of a "huge expansion factor," he doesn't mean that it will be two or three times as long. He means that the written transcription of what seems obvious to us might be ten, a hundred, or a thousand times as long as the summary we make for ourselves in our head. And even then you'll have to leave aside a bunch of details you won't have the heart to write down.

The phenomenon Thurston describes isn't limited to high-level research. It shows up as soon as we try to faithfully describe even the simplest of our mental images. An image is worth a thousand words, and unfortunately that also applies to images that exist only in our heads.

To get a better idea of this, let's pick up one of our favorite examples. How much time does it take you to visualize what you do when you tie your shoes? Two seconds? Three? Now take pen and paper and try to describe each movement exactly, so that an absolute beginner could follow your instructions and get the same result. There is a difficult version of this exercise where only words are allowed. But the easier version, where you can use drawings, is already immensely hard.

Once you grasp the degree of this difficulty, you'll understand something fundamental and profoundly comforting: a math book might seem to be terribly complicated even though the ideas are quite simple.

There's really no reason to be afraid of mathematical writings. Not only is the reading between the lines Thurston talks about possible, but it's necessarily simpler than the text itself.

But before getting to this simple understanding, while you still don't have the right mental images, you'll have to go through a lot of trial and error.

The Art of Clarity

It's not surprising if you couldn't find the words to describe how you tie your shoes. Most likely, you didn't even try. Writing mathematics, that is, transcribing mental images with enough clarity and precision to allow others to understand and reproduce them, is an art.

What makes it so difficult is that your mental images are often a lot less clear than you think. What keeps the knot you make in your shoelaces from coming undone? If you don't know, you don't really know what it takes to tie it right.

As Grothendieck explained, the art of mathematical writing is really a dual task of the clarification of ideas and the refinement of language. It's a delicate exercise of motor coordination and requires years of practice to master. The good news is that with patience and effort anyone has the ability to get better.

Learning to write math is learning to have clear ideas. Wouldn't it be a shame to deprive yourself of that?

By writing math yourself, you'll get to understand why it is written in such a bizarre formalism, in this language made for robots: there's really no other choice.

To demonstrate, we'll return to another of our favorite examples: the concept of shapes, such as you discovered in your early childhood. Imagine an alternative universe where you were the first person to discover the concept of shapes. How would you explain in words your method of distinguishing stars from squares and putting the right blocks in the right holes?

The Patience Game

In this alternative universe the visual culture is so impoverished that the *shape game* is called the *patience game,* because the only known method to solve the problem is through hours of trial and error.

There's no geometric language. There's no word for "round," "square," or "triangle." The word *heart* is used only for the muscle that beats in our chest. If you used this word to point out one of the pieces in the patience game, no one would understand you. The same thing with the word *star.* Stars shine in the night sky but no one would see the relation to the patience game. The question's not even

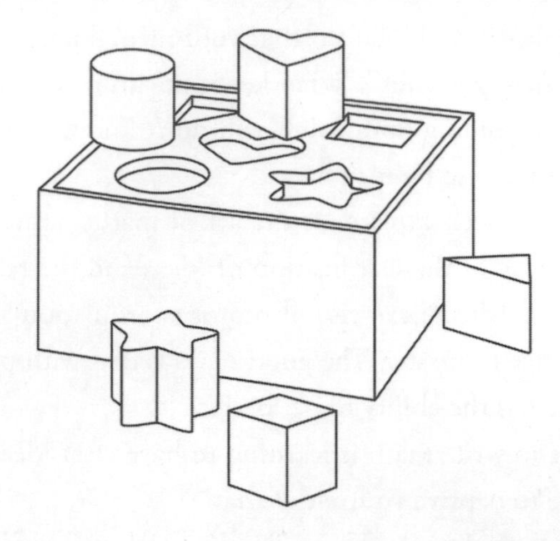

whether the stars have five, six, or seven points. People are far from that. And besides, where do you get the idea that stars should have *points?* What does that even mean?

In your way of seeing the world, in your inner language, you accept the idea that stars have points and you recognize a five-pointed star as one of the blocks in the patience game. Why not? Except that so far this idea exists only in your head.

The others aren't blind. They are biologically capable of seeing the same shapes as you. But they haven't yet learned how. Their brain receives the same raw visual information but doesn't structure it in the same way.

"Look, this block is in the shape of a star. There's a hole also in the shape of a star. If you take this block and put it in this hole, the right way, it goes in on the first try."

This kind of explanation doesn't work. People can't see the star that's in front of their eyes. They may very well live in the same world as you, but their mental experiences are different. They'd laugh out loud if they saw you solve the game without trial and error. You'd look like a magician.

The Theory of Touch

Instead of geometry, the inhabitants of the alternative universe have a highly developed sense of touch. Schoolchildren are all taught the *theory of touch*. They learn how to run their fingers along surfaces and recognize textures. They learn soft, hard, smooth, rough, grooved, stringy, raspy, friable, porous.

When they hold a block, they readily identify some protruding edges (what they call *points*) and also some concave grooves where their fingers get stuck (these are called *pits*).

It might not be much, but it's a starting point. The power of math lies in its ability to extend language with new words that are precisely defined. By relying on things people already understand, you construct new ones that they can't immediately feel but that they can nevertheless manipulate through definitions.

By playing around with these new words, hopefully they'll end up someday really understanding them.

To talk about triangles, stars, and squares without relying on the language of vision, you have to reconstruct these concepts by relying on the vocabulary of touch. You can't simply point your finger and say, "Look, that's a star." This inability to rely on a shared experience will pose tremendous difficulties in writing and you'll end up with a complicated and unfathomable text. But it's doable.

This is what the result might look like. Careful: the following few pages are written in a style that greatly resembles official math. Consequently, they are quite frankly a pain to read.

Tactile Theory of the Patience Game, for Beginners

When you run your finger along the edge of a block (or hole), you come across a series of points and pits. Let's give a name to this series of points and pits, and call it the *signature* of the block (or hole).

For example, there's a block (the one you want to call triangle, except that this word doesn't yet exist) whose signature is:

point, point, point

and a hole (the one that fits the block) whose signature is:

pit, pit, pit.

As always with mathematical definitions, the name is arbitrary. I chose "signature," but I could have chosen a different name and it wouldn't change anything, since the meaning I assign to the word is entirely contained in the definition, without any direct relation to its everyday meaning. But since I have the choice, I may as well pick a word that helps readers, a word whose everyday meaning might help them understand the mathematical one. *Signature* seems good to me since it evokes the idea that the signature lets you identify individual blocks and holes. If this word doesn't work for you, feel free to use another.

The definition I've given, however, has a minor technical problem: the same object can have a number of different signatures, depending on where you start. To be rigorous, therefore, we should talk about "a" signature rather than "the" signature. For example, take a block (the one you want to call a star) whose signature is:

point, pit, point, pit, point, pit, point, pit, point, pit.

But if you start your finger at a different place you could also come up with this signature:

pit, point, pit, point, pit, point, pit, point, pit, point.

What matters isn't the signature itself but the *signature up to rotation.* But this notion needs to be properly defined. To do this, I first need an intermediate notion: the *elementary rotation* of a signature is defined as the signature obtained by taking the first word and putting it at the end. For example, the elementary rotation of

pit, point, point, point

is

point, point, point, pit.

The two signatures are *equivalent up to rotation* if you can go from one to the other by a series of elementary rotations. For example, the four following signatures are equivalent up to rotation:

pit, point, point, point
point, point, point, pit
point, point, pit, point
point, pit, point, point.

Each line is obtained from the previous one by applying an elementary rotation. If you apply yet another elementary rotation to the last line, you'll get back to the first.

Definition. A *shape* is an equivalence class of signatures up to rotation.

To make sense of this definition, you need to know what an *equivalence class* is. It's a basic mathematical concept, and you'll find the definition in any set theory textbook. In practice, it means any signature defines a form and that two signatures define the same form if and only if they are equivalent up to rotation. Taken together, the four signatures above constitute an example of an equivalence class of signatures up to rotation.

If I wanted to, I could continue to invent words. I could define triangles, circles, and squares in terms of their signatures. For example, a triangle would be defined as a shape whose signature is:

point, point, point.

In the same way I could choose to define a *star with n points* as the form whose signature is obtained by repeating *n* times the formula *point, pit.* In the particular case of the five-pointed star, the signature would be:

point, pit, point, pit, point, pit, point, pit, point, pit.

A *heart,* by definition, is a star with a single point.

The language of signatures and shapes will enable us to describe a solution to the patience game that eliminates the need for patience.

Definition. The *mirror image* of a signature is the series of words obtained from the signature by systematically replacing the word *pit* with the word *point* and the word *point* with the word *pit.*

Thus the mirror of *point, point, point* is *pit, pit, pit* and vice versa. If the two signatures are equivalent up to rotation, their mirror images are as well, and so the concept of mirror images extends to shapes. The main result of the theory of the patience game is as follows:

Theorem. For each block B there exists a unique hole H such that the shape of H is the mirror image of the shape of B, and H is the only hole B can fit into.

This can be turned into a simple method to determine which block fits in which hole:

1. Run your finger around a block to determine its shape.
2. Run your finger around each hole until you recognize the corresponding mirror shape.
3. Once the mirror shape is found, you know you have the right hole, and you can fit the block into it.

Real Pleasure

As with all mathematical definitions, our definition of shapes may seem arbitrary and unwieldy.

The good news is that we've managed to talk about shapes using the vocabulary of tactile experience, without any reference to visual perception. In other words, we've found a way to describe the shape of a star in a language a person without sight could understand. This undoubtedly is a tremendous feat.

The bad news is that our definition is ugly. It absolutely fails to capture the richness and beauty of the visual experience, and all that shapes represent for us: their obviousness, their universality, their unescapable presence, and everything that makes us love them. For all this our definition is a very poor substitute.

But it would be stupid to think we've finished. We've only just begun. To unravel the ball of yarn of what a shape really is, our definition is a poor beginning among innumerable other possible beginnings. Nothing prevents us from working harder and crafting a language that will capture with more and more finesse what it means for a star to be more or less pointed, more or less elongated, more or less distorted, and so on.

Yet extending the vocabulary, adding precision and detail, won't fix the issue. Our problem runs deeper. Seeing isn't a question of words.

Seeing is a sensory, instinctive experience that we live without having to think about.

Saying that a shape is an *equivalence class of signatures up to rotation*, saying that a star is *a shape whose signature is obtained by repeating* n *times the pattern "point, pit,"* is clever enough, but it can never be entirely satisfactory. We're not robots. We have no wish to apprehend the world through a language that seems to have been invented by dystopian bureaucrats. We want to "see" without thinking about it.

When you puzzle over a text written in the official language of mathematics, you're in the situation of a sightless person who is working out our formal definition of stars without being able to see them. It's absolute gibberish, at least at first, as long as you're unable to get an intuitive sense of the definition and access "the thoughts between the lines."

Mathematical comprehension is precisely this: finding the means of creating within yourself the right mental images in place of a formal definition, to turn this definition into something intuitive, to "feel" what it is really talking about.

Understanding a mathematical text that defines a star as *a shape whose signature is obtained by repeating* n *times the pattern "point, pit,"* is to get to a point where you forget the formal definition and sense directly what a star is, on command, with the simple mention of the word *star.*

The real pleasure of mathematics is waking up one day and realizing that you can see stars in your head, which you'd never been able to do before.

The secret techniques of mathematicians aim to facilitate and accelerate this intuitive understanding. Mathematicians use logic and language as an apparatus for *learning to see.*

I know that seems too good to be true, and it probably feels way out of your league. That is, however, not the case. You have the ability to start with an abstract definition and intuitively sense what it designates. You've already done this.

No one has ever described to you, other than through a complex assemblage of abstract mathematical concepts, the number that you can picture in your head by taking a billion and subtracting one.

Its written form as 999,999,999, which gives you the impression that the number is physically present on the page in front of you, is just the shorthand notation for an incredibly complex formal definition. It characterizes this number as the result of a chain of additions and multiplications that, if you tried to imagine it, would give you a headache:

> *Nine plus nine times ten plus nine times ten times ten plus nine times ten times ten times ten plus nine times ten times ten times ten times ten plus nine times ten times ten times ten times ten times ten plus nine times ten times ten times ten times ten times ten times ten plus nine times ten times ten times ten times ten times ten times ten times ten plus nine times ten times ten times ten times ten times ten times ten times ten times ten times ten.*

On paper, this number is an abstract assemblage, logical and cold. Yet in your head, it's a simple object, concrete and clear as day.

9

Something's Going on Here

During my school years, I was often told that I wasn't holding my pen correctly, and that's why I wrote like a pig.

I chose to study math because I thought that they'd teach me how to "hold" math correctly in my head. I did it my own way and that worked well enough, but I was never at all certain that my way was the right one.

My biggest surprise, throughout my studies and scientific career, was never having received any formal education in the subject, as if it weren't serious or worth the time.

I was, no doubt, a bit naïve, but it seemed to me that the central problem in math wasn't whether such and such theorem was true, but why it was so easy for some and so difficult for others. After high school, in my first year as an undergrad, I was expecting that the first class would focus on the right way to mentally manipulate mathematical concepts. Surely they'd begin by explaining how to do it!

But the first class was on a different subject. In official math, the starting point isn't the unseen actions that you do in your head, it's formal logic and set theory. The explanation I was waiting for didn't come in the next class, or the one after that. In the end I gave up waiting for it.

The question, however, came up again a few weeks later, as we began the study of vector spaces in arbitrary dimensions. That's when I became seriously preoccupied with it.

A one-dimensional vector space is just a line. A two-dimensional vector space is a plane. We live in a three-dimensional vector space, or rather we have the tendency to believe we do, even if Einstein showed us how that's not entirely true.

There's no reason to stop at three. With logical formalism, you can keep going. You can define what a four-dimensional space is, five-dimensional, six-dimensional, and so on. If you wanted to, you could do geometry in 24 dimensions, or 196,883 dimensions, or dimension *n,* where *n* is any whole number.

These spaces aren't laboratory curiosities. They are fundamental concepts, without which we can't understand the world around us, and they are so central to modern science and technology that for more than a century they've become part of the basic vocabulary, like whole numbers.

If you've never learned to think in multiple dimensions, you've missed out on one of the great joys of life. It's like you've never seen the ocean, or never eaten chocolate.

The Fruit of Your Imagination

When you do geometry in two or three dimensions, there's an easy way to show what you're talking about: make a drawing. For example, in a three-dimensional space, you can put together twenty equilateral triangles to make a twenty-sided die that looks like that pictured in the figure.

This remarkable object, which has been known for ages, is called a *regular icosahedron.* Looking at the drawing, you have the impression of seeing an icosahedron floating in space. But that's not really what's in front of your eyes. What you're looking at is a two-dimensional page on which is found the image of an icosahedron. More precisely, this image is what is called a *projection:* it's the shadow (in two dimensions) of an imaginary icosahedron (in three dimensions).

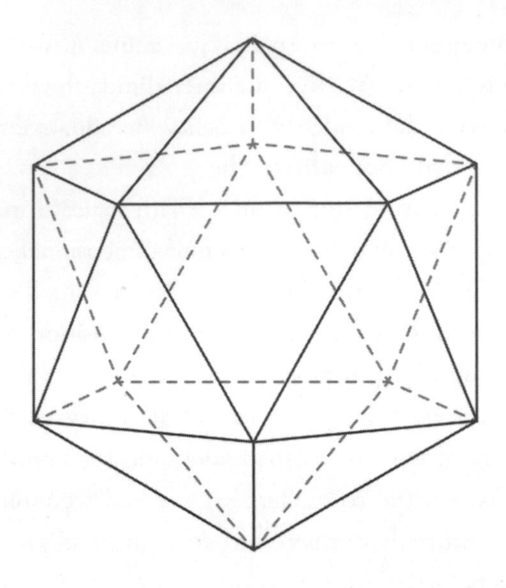

Your mind easily reconstructs three-dimensional things from their two-dimensional projections. When you look at your vacation photos, you have the sense of actually seeing scenes that occur in three dimensions. It doesn't require any particular effort. It doesn't tire you out or raise any metaphysical concerns. You never believe that the scenes are actually taking place in two dimensions. You never get the impression that your three-dimensional perception is nothing but an abstraction, a mental reconstruction that is simply the fruit of your imagination. You never have the sense that what you see in the photos is a hallucination.

Your mind can even see things the image doesn't show. Looking at the projection of the icosahedron, not only can you see it, but you can turn it around in your head, even if that requires a bit of concentration. The actual drawing stays perfectly still. But that doesn't stop you from easily understanding what I mean by "turning" the icosahedron in your head.

For example, if you turn the icosahedron a fifth of the way around its vertical axis, you'll find the same icosahedron you started with. Rotational invariance is a well-known property of icosahedrons.

If I had simply defined a regular icosahedron as an abstract assemblage of twenty equilateral triangles, without having given you the means to picture it, you would have had a lot more difficulty understanding this rotational invariance. With the drawing, it's a lot easier.

Visual intuition makes certain mathematical properties clear, that without the mental image wouldn't be clear at all. This is why transforming mathematical definitions into mental images is so important. When you're unable to imagine mathematical objects, you have the sense that you don't really understand them. And you'd be right.

Geometry for the Sightless

When you hear someone talk about four-dimensional geometry for the first time, you ask yourself what this fourth dimension could be. Is it time? Something else?

The right answer is that the fourth dimension is whatever you want it to be.

When you do geometry in two dimensions, on a plane, a point is determined by two coordinates, generally called x and y, that represent exactly what you want them to represent.

—When you look at a map, x is generally the longitude and y is the latitude.

—When you draw the façade of a building, x is generally the length and y the height.

—When you represent the growth of a population of rabbits, x is generally time and y the number of rabbits.

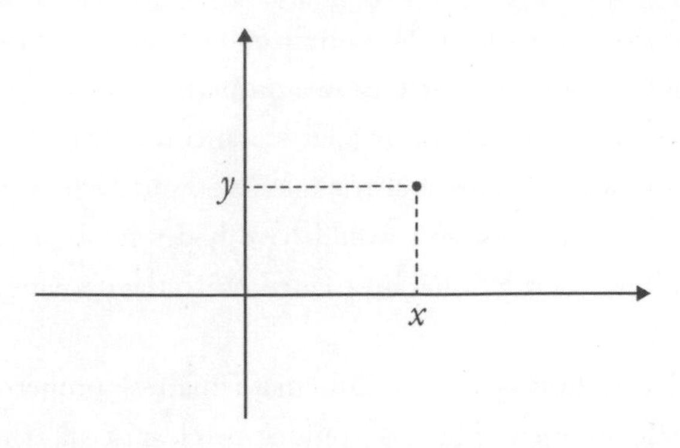

In the same way, in a space with ten dimensions, a point is determined by ten coordinates that are generally called $x_1, \ldots, x_{10}$. If you want these coordinates to represent something, it can be whatever you want.

If you wanted to describe the geographic expansion of an invasive population of rabbits, you would need to think in four dimensions, since you'd need four coordinates: *longitude, latitude, time,* and *population density.*

It's correct to say that four-dimensional geometry is an abstraction. But it's a simple and natural one. Your mind can accept the fourth dimension and even find it concrete, the same way that it accepts as concrete all those things that in reality aren't at all concrete. The *geographic expansion of an invasive population of rabbits* is an abstract notion. If it seems concrete, it's because your mind has already accepted the idea that the fourth dimension exists and that it's concrete.

Contrary to common belief, it's never abstraction that makes math difficult to understand. Abstraction is our universal mode of thinking. The words that we use are all abstractions. Speaking, making sentences, is to manipulate and assemble abstractions. In that respect, four-dimensional geometry isn't any more abstract than two-dimensional geometry. The problem with four-dimensional geometry has nothing to do with abstraction. The problem is that it's hard to visualize and hard to draw.

Geometry courses on higher dimensions are geometry courses for the sightless. They resemble the theory of touch discussed in the previous chapter: instead of relying on visual cues, they use mathematical language and formalism to define a geometric vocabulary whose meaning is very precise but whose visual interpretation isn't straightforward. Everything can be described using formulas based on coordinates. For example, there is a formula that defines the distance between two points from their coordinates.

At first, our minds aren't used to this new vocabulary. We can't give it an intuitive visual meaning. This is why four-dimensional geometry can't be taught in the same way as two-dimensional geometry, where figures and direct visual intuition play a central role.

For example, there's a four-dimensional analogue to the icosahedron. It is a die with 600 sides, insanely regular and even more beautiful than the icosahedron. Or rather, you should say that this object is a "hyper-die" with 600 "hyper-sides." These hyper-sides are three-dimensional objects that are *regular tetrahedrons,* that is, regular pyramids with a triangular base. Each hyper-side has 4 sides that are equilateral triangles, along which it is attached to 4 other hyper-sides. In total there are 600 hyper-sides, 1,200 sides, 720 edges (the sides of the triangles) and 120 vertices.

Are you having problems visualizing it? If it's any help, I'm providing a drawing.

The picture is the two-dimensional shadow of the four-dimensional hyper-die (or rather *one of its shadows,* since an object's shadow changes according to its orientation toward the direction of the light).

It would be agreeable to look at this image and, without any ef-

fort, see the "hyper-icosahedron" floating in front of you, suspended in four-dimensional space.

I'd also like to see it floating before my eyes. I'd like to grasp it at a glance and perceive its multidimensional thickness. Sadly, that's not what happens.

My mind isn't capable of instantly and effortlessly constructing a mental image of a four-dimensional object from its two-dimensional shadow. I do manage to perceive the physical presence of the hyper-icosahedron, but I do it in a different way, without relying on the drawing.

Truly False Images

In studying math, I quickly realized that I was just like everyone else, and I didn't know how to see higher-dimensional geometric objects the same way I could in two or three dimensions.

But I also noticed another phenomenon, quite subtle and unexpected.

This phenomenon wasn't anything special. It was going on behind the scenes, in the background, and could easily have gone unnoticed. Maybe it had been going on forever without my even noticing.

At any rate, it was only at this very moment, a few weeks before my eighteenth birthday, as we had begun looking at higher-dimensional geometry, that I became aware of it: the abstract geometric notions that I had been taught evoked in me bizarre impressions that were more or less visual.

These impressions were very faint and their meaning was unclear. They were fleeting mental images: confused, vague, and evanescent. They were also unstable. Sometimes they were there, sometimes not. When they were there, the images were always naïve. And what's worse, they were always false.

It was as if my mind were trying to see higher-dimensional geometry by cobbling together two- and three-dimensional mental images. The result was ridiculously far off the mark. The images weren't just a little false, like a circle that's drawn is false because it's not perfectly round. My mental images were *grotesquely* false.

I was studying at Louis-le-Grand, the most prestigious preparatory school in Paris. I was being taught serious math, the official mathematics with all its axioms, definitions, propositions, theorems, proofs, symbols, and formulas. The teaching was logical and structured. I was taught to write math rigorously and precisely.

And yet, for some reason, my naïve intuition hadn't entirely given up. It was as if it was refusing to sink into irrelevance. It didn't work at all and gave bizarre results.

The designs in my head resembled preschool scribblings, like when I drew people with their arms and legs attached directly to the head, without realizing I was forgetting a significant part of the body. Or rather realizing, in a confused and mixed-up manner, that I was forgetting something of importance, without figuring out what it might have been. Something was wrong, I knew, it but I couldn't say what it was.

I have this clear memory from when I was four or five. One day, I called on the teacher to tell her there was an issue with my drawing. Something going's on here, but what? She replied that everything was fine, that my drawing was actually very nice. I had the very distinct impression she was making fun of me.

I had no desire to replay this experience and raise my hand to say that I had a problem because the images in my head were false. I had no desire to be ridiculed in front of everyone. My reflex was to treat these false mental images like parasitical thoughts that I had to get rid of.

If the secret of success in the preparatory class at Louis-le-Grand was to think like a four-year-old and make scribbles in your head, everyone would be onto it.

It made no sense to insist on seeing things simply and in images. It was time to grow up and learn how to think logically, in an organized manner, using complex and serious words. I needed to act like an adult.

Bigger or Smaller Pipes

At the time, I still believed that logic helped you to think. I wasn't myself able to think logically, but I thought it was because something was wrong with me. I believed that studying math would help me fix this issue, and that the first step was to get rid of my naïve and false mental images.

But amid all these false images, amid these parasitical thoughts that I wanted to get rid of, I was surprised to find one that was less false than the others.

When you study vector spaces, you also study the concepts of *linear maps, kernel, rank, dimension,* and *codimension.* Vector spaces are usually represented by letters and linear maps by arrows connecting these letters. But when I chose to picture vector spaces as bigger or smaller barrels (according to their dimension) and linear maps as bigger or smaller pipes (according to their rank) connecting those barrels, then all the exercises on these concepts suddenly became obvious.

It wasn't much. That was a tiny subject and there were plenty of other exercises that I was unable to solve. But in this subject, not only was I able to solve the exercises, they became as obvious to me as 1,000,000,000 – 1 = 999,999,999. They became so obvious that it seemed absurd that you'd even have to ask, and even more absurd that there were people who didn't know how to solve them.

These barrels and pipes made my life simpler, but where did they come from? Was that how you're supposed to do it? What was going on in other people's heads? How were they imagining mathematical concepts?

I remember looking perplexedly at my classmates, scrutinizing their faces to look for signs of what was going on inside their heads.

I was shocked to find out I hadn't the slightest idea.

A Gigantic Problem

No one had told us what we were supposed to be doing inside our heads, and that was becoming a gigantic problem.

I realized that there were two radically different ways of looking at the instruction we were receiving, and that the two approaches were mutually incompatible.

The first approach consists of treating mathematics as knowledge. Mathematical statements are information that you have to ingest and be able to recall. You have to learn the definitions, learn the theorems, learn the proofs.

The second approach consists of refusing to learn. It takes on math as a sensory experience. The sole function of mathematical statements is to help you generate mental images, and only these images will lead to comprehension. Once you have the correct mental images, everything else becomes clear.

The two approaches are incompatible because they require opposite mindsets. Learning by heart, accepting to believe what you don't understand—that only exists in the first approach. In the second approach, you look at what you don't understand with distrust and incredulity: "Really? This is supposed to be true? No way! How's that possible? How can I picture that?"

Until then, I had instinctively followed the second approach, and it had worked rather well for me. In primary school, when the teacher explained what a circle was, I knew beforehand how to picture circles in my head. It's why math had been so easy for me. School had taught me how to put words to things that were already easy for me to picture more or less clearly.

But I was reaching the end of that road. I was now being introduced to serious and profound things that I couldn't understand and that my intuition was incapable of visualizing. I was reaching the limits of my mental capacity. The concept of these bigger or smaller pipes was perhaps my final valid intuition. And even that was a matter of luck. With such naïve images, frankly, what more could I expect?

And without being able to rely on my intuition, I was stuck. I didn't have any choice. The moment had come when I finally had to "learn."

But I was well aware of the implications. Treating math as knowledge meant giving up the pleasure of understanding it, of feeling it come alive within me. It meant giving up loving it.

An Ear for Dissonance

To be honest, it wasn't the first time math posed real difficulties for me.

It had already happened at the beginning of middle school, when we had to use letters to designate numbers. "Let n be a whole number." But if n was a whole number, why not just say the number? Why all this mystery? Couldn't they just spit it out? I had the feeling of being left out of the game, of not understanding anything, not being smart enough.

At the time, thankfully, this didn't last long. After a few weeks, without any conscious effort, I ended up accepting the idea that you could reason with letters, that is, reason with numbers without knowing the actual number. I understood that not knowing the actual number was in fact the whole point. Reasoning with letters was a way of reasoning with all numbers at once. It was doing an infinite number of computations with a finite number of words.

At Louis-le-Grand, however, things were going much faster, and the situation looked unsurmountable. Each week there were ten new concepts to learn and I had no idea how to arrange them in my head.

It was at this very moment, right before I turned eighteen, that I made the most significant decision in my scientific career, and possibly of my entire life: instead of ignoring my stupid ideas and parasitical thoughts, I decided to embrace them. I chose to listen to them and take them seriously.

Of course, that didn't mean taking them at face value. I knew perfectly well that they were false. It was obvious. But since it was so obvious, was I able to say *exactly* in which way they were false?

Today, when I try to describe this intellectual method, I sum it up like this: I began to listen to the dissonance between my intuition and logic. In chapter 11, I'll explain what that means in practical terms with an example that should speak to you.

In hindsight, I'm amazed that I made this decision on my own, piecing things together myself without anyone to tell me it was the right way of doing things.

I remember trying to speak about it with a friend, Xavier, who was in the same class. My concerns were so far off from official math and what we were being taught that I failed to express myself clearly. I didn't have the right words to talk about these issues. It took me decades to find a proper way to talk about them.

I had no reason to believe that this method would work and I wasn't even hoping that it would. It was just a childish experiment. I was carrying it out through sheer curiosity, just to see, to find out exactly when and how it would fall apart. I was expecting it to fail, because it seemed unthinkable that there was a method for understanding math and we had been kept in the dark.

However, I soon found out that my approach worked. The more I thought about my stupid ideas, the less stupid they became. The more I focused on my parasitical thoughts, the clearer they became. The more I listened to the dissonance between my intuition and logic, the more I was able to transcribe it in words. My intuition was never perfect, but it kept progressing without any effort on my part.

In the space of a few weeks, my way of studying was transformed. I began to use class as a benchmark for my intuition. I tried to predict what the teacher was going to say. Most of the time I got it wrong, but that let me figure out where my intuition was already correct. The things I understood, I understood so well that I could rely on them and concentrate on the others.

I kept going back to what I didn't understand until I understood why I didn't understand it. And in the end that's what allowed me to understand.

Our Ordinary Intuition

As long as schools refuse to teach the human reality of mathematics, all mathematicians will be self-taught.

Saying that math is a matter of intuition isn't enough. You also have to explain that this intuition is accessible and describe the ways that you can develop it. Nothing is more intimidating than the myth that mathematical intuition is something special that only a chosen few are endowed with.

Mathematical intuition is the same intuition we use every day, but developed and solidified by its confrontation with language and logic. It's what our intuition becomes once we stop believing it's a gift from heaven, and instead work on improving it.

Math often feels like gardening. You weed, plant, prune, water. It seems at first that nothing is growing, yet one day you realize that it has. It's hard to believe that you can start with our common perception of space and develop it so that it becomes instinctive to think in whatever dimension. However, that's the case.

Behind our false beliefs about mathematical intelligence, behind our superstitions and inhibitions, there lies our ignorance about mental plasticity and the laws that govern it. We'll come back to that in the next chapter.

Why math education consistently misses that point remains a mystery to me. It's as if teachers felt they're not qualified to talk about it. Very few people dare to describe their intuition in simple language and admit to the great naïveté of their mental images.

Nevertheless, some of the greatest mathematicians have done so with disconcerting tranquility. Pierre Deligne is a striking example. The most successful of Grothendieck's students and himself an extraordinary mathematician (Grothendieck used to say of Deligne that "he's better than me"), Deligne won the Fields Medal in 1978 (for solving the famous Weil conjectures) and the Abel Prize in 2013.

In his 2013 Abel Prize interview, Deligne was asked about his working style. That was the occasion to talk about his intuition and his perception of higher-dimensional geometry. Here, among other things, is what he said:

> It's important to be able to guess what is true, what is false. . . .
>
> Then I don't remember statements which are proved. I try to have a collection of pictures in my mind. More than one picture, all false but in different ways, and I know in which way they are false. . . .
>
> The pictures are very simple. I draw just in my mind something like a circle in the plane and a moving line which sweeps it, but then I know that this is false, that it's not one-dimensional but higher dimensional. . . .
>
> It's always very simple pictures, put together.

Mathematical intuition is so banal, simple, and stupid, that you need a lot of self-confidence not to throw it in the trash. When you're no longer a child, you have only one wish, and that's to reduce your false intuitions to silence. That's what almost happened to me when I thought I had to get rid of my stupid ideas and parasitical thoughts.

The shy little voice that's telling you that you don't understand, that's your mathematical intuition. Don't confuse it with the loud noisy voice that's telling you that you're not smart enough. The little voice will guide you. That's the one you need to lend an ear to. That's the one you need to take care of, and protect throughout your entire life.

10
The Art of Seeing

When I think about higher-dimensional geometric objects, I have a clear enough visual perception. But I don't see them in the same way I see objects in the physical world. I really only see certain aspects, certain pieces, the details that interest me. I'm not really able to see them entirely. But I sense their presence in a diffuse way, with my entire body.

Being able to see in four or five dimensions in the same way we see in three dimensions is said to be impossible.

Nevertheless, Bill Thurston could do it. This amazing ability is exceedingly rare. It elicits admiration, even among the mathematical community, where it contributed to his legend.

It's natural to find that intimidating. It's tempting to see in it the proof that the great mathematicians are different, with a brain that is biologically superior to our own. However, when you get to know Thurston's personal history, you realize that this hypothesis of an extraordinary gift doesn't hold. His story isn't that of an alien endowed at birth with the extraordinary ability to see the world in five dimensions.

In fact it's rather the opposite. His story is that of a young boy born with a handicap that during the first years of his life kept him from seeing the world in three dimensions.

Thurston was born with a strong congenital strabismus, or squint. The visual fields of his two eyes never met. When he looked at an

object, he could see it with only one eye at a time. The two images couldn't come together, which deprived him of any direct perception of depth.

He was fortunate to have a generous and loving mother, who from the time he was born devoted a lot of time and energy to help him overcome his handicap. When Bill was two years old she worked with him for hours, showing him special books filled with colors and designs, pretexts for long exercises of reeducation.

Thurston's almost carnal love for geometry, a love that would continue throughout his life, traces back to this period. This love of space, materials, textures, and forms underpins the entirety of his work and is reflected in the marvelous drawings that illustrate his manuscripts.

When he began primary school, Thurston resolved to work each day to expand his capacity for visualization. Very early in his life, without knowing it, he became an extraordinary mathematician.

It would be a mistake to believe that he spent more time than other children learning how to see. Every second we have our eyes open, as we observe the world around us, we're building on our capacity for visualization. Learning to see is one of the main activities of the first years of our lives (and not only the first years). But for the vast majority of children it's an unconscious activity. It happens continually, in the background, almost unintentionally, and without much concentration.

Thurston didn't have this luxury. He couldn't simply let nature take its course. For him, there was nothing instinctive about seeing the world. Learning to see was a conscious project. You could even say it became the work of a lifetime.

Fosbury wasn't able to jump like others so he had to invent his own way of jumping. Thurston wasn't able to see like others so he had to invent his own way of seeing. Fosbury's approach allowed him to jump higher. Deliberately working to see better, as Thurston did, allows you to see further, more distinctly, down to the heart of things.

When we believe we're directly seeing the world in three dimensions, we're unconsciously piecing together the two-dimensional images captured by our retinas. This way of perceiving space is imperfect. It's not an objective perception but a subjective one, relative to the place we're looking from and distorted by the effect of perspective. What's worse, from our local perspective, the greater part of the world is hidden to us.

Thurston was denied this easy access to three-dimensional perception. He worked to construct his own, in his own way, through the power of thought. If Thurston had a gift, it was that of patience and determination. Or perhaps that of love and self-confidence.

Mathematical work isn't a series of lightning insights and strokes of genius. It's first of all a work of reeducation based on the repetition of the same exercises of imagination.

Progress is slow because the body needs time to transform itself. It doesn't help to force it, which may end up hurting you. You just need to commit to a regular training schedule, keep your cool, keep going even when it seems you're not making any progress. It's like going to the speech or physical therapist, except you're all alone and inside your head.

Seeing Further

Thurston consciously and conscientiously developed his ability to picture the world. Through persistence, by working to stitch together two-dimensional images in his head, he was able to learn to see in three dimensions.

But why stop there? He realized that with the same technique he could go even further. By putting together images in three dimensions, he learned how to see in four. And by putting together images in four dimensions, he learned how to see in five.

Even in only three dimensions, Thurston's approach made him able to see things no one had seen before. His geometrization conjecture, formulated in 1982, deals precisely with the third dimension. A conjecture is a mathematical statement that someone believes is valid but isn't yet able to prove. Making a conjecture is feeling something is right without being able to say why. It is by nature a visionary and intuitive act.

Thurston's conjecture was a spectacular breakthrough. It notably encompassed the famous Poincaré conjecture, which had been formulated in 1904 and remained unsolved for so long that it was designated a "Millennium Prize Problem" in a list established in 2000 of seven mathematical problems judged the most profound and difficult, each with a million-dollar award for its solution.

In 2003 Grisha Perelman was able to prove Thurston's conjecture. He was thus also able to solve Poincaré's conjecture. We'll talk more about Perelman and the million dollars in chapter 17.

Seeing Is Believing

When Thurston claimed to see in four or five dimensions, what did he mean? What was he seeing exactly, and in what sense should we understand his use of the verb *to see?*

The best way to answer these questions is to turn them around. What exactly do *we* see? What do *we* mean by "see"?

In using the verb *to see* for ourselves, we have the tendency to overestimate its meaning. When we look around us, we have the illusion of a direct relation to the world, as if our eyes were magic windows drilled directly into our consciousness and giving us direct access to reality. If that's the meaning we want to give to the verb *to see,* then we have to be ready to face the consequences: in this case, we never really *see* anything, we just *believe* that we do.

What you see is never really unvarnished reality but an interpretation of the world. In other words, it's a reconstruction, produced by your memory and imagination, based on raw visual signals that you're never directly aware of. The day you were born, you didn't yet know how to see, because your brain hadn't yet learned to give a sense to the raw information delivered by your optic nerves.

This same ability for reconstruction allows you now to imagine things that don't exist, and to have the impression of seeing them. Seeing something and imagining seeing it aren't that different. You can imagine seeing an ant as big as a horse, right there, before your eyes. You can picture it, describe it, and even say a lot of things about it. But others have no means of directly seeing the giant ant in your head.

It's the same with colors. Red gives you a very particular sensation. But what exactly is the sensation of red in your head? How do you know it's the same as mine? Maybe that question itself makes no sense.

Dalton and His Geranium

We have such a hard time describing and communicating our visual sensations that it wasn't until the autumn of 1792 that someone realized that we aren't all equal in the biological perception of colors: around 8 percent of men (and 0.6 percent of women) are color-blind.

How is it that such a striking and easily shown fact could remain unknown for thousands of years, from time immemorial when our ancestors began to speak of color?

It was John Dalton, the great physicist to whom we owe the modern idea that matter is composed of atoms, who made the discovery based on his own case. He revealed his findings in a scientific communication whose sensationalist title betrays his own personal stupefaction: "Extraordinary Facts relating to the Vision of Colors."

Dalton seems almost as astounded as if he had been the first to discover that there were right-handed and left-handed people. But when you read his account, you understand the mechanism that allowed this misunderstanding to persist for millennia.

From the point of view of modern science, the story is simple enough. Our perception of colors is explained by the presence within our retinas of dedicated cells called *cones.* A human eye normally has three types of cones, sensitive respectively to blue, green, and red light. We perceive a large palette of nuances of colors, but only through their relative proportion of blue, green, and red. (It's for this simple reason that screens combine these three colors in each pixel.) Dalton was a carrier of a genetic mutation. He only had two types of cones. He was missing the cones sensitive to green light, which kept him from perceiving certain nuances. For example, he found it hard to distinguish blue from pink. But Dalton grew up in a world where no one imagined that was possible, so he learned more or less how to name all the colors that others saw and talked about. He convinced himself he was really seeing them. He saw the world in color, without ever suspecting that he was missing something.

However, when told from Dalton's point of view, before words were invented that allowed for its scientific explanation, his story reads like an absurdist comedy.

Dalton begins his account with a confession: all his life, he'd had the impression that the names of colors weren't well chosen. When people sometimes used the word *red* instead of *pink,* Dalton found it absurd. Pink, for him, looked more like blue, nothing at all like red. But he never dared to tell anyone.

In 1790, Dalton began to get interested in botany. He had a hard time recognizing the colors of flowers but that didn't surprise him too much. He simply got help. When he asked people whether a flower was blue or pink, he saw in their eyes that they thought this had to be some kind of prank. He didn't understand why they looked puzzled, but never took the time to find out. He was already accustomed to the strangeness in all conversations about color. The misunderstanding could have gone on forever if Dalton hadn't made the discovery, in the autumn of 1792, of a geranium with absolutely extraordinary properties.

This geranium was said to be pink. However, in sunlight, it looked blue. So far, nothing unusual: for Dalton, those two colors were always very close. But when Dalton had the idea to look at it by candlelight, the geranium turned bright red, a color that for him had nothing to do with pink.

Dalton was so flabbergasted that he brought his friends over to admire his miraculous geranium. When his friends told him that his geranium was nothing special, he was crestfallen. The only person who seemed to understand was his own brother. That was the starting point for Dalton's experiments on the perception of colors that allowed him to demonstrate color blindness (sometimes called *Daltonism*) and its hereditary character.

There are three morals to this "extraordinary" story.

The first is this: between raw perception and what we believe we see, there's a lot of room for maneuver. Dalton found the names of colors strange but that didn't keep him from accepting them and interpreting them in his own way. He constructed his own color scale, different from non-color-blind people's, but not necessarily more impoverished. The most striking example was his hyper-sensitivity to the nuances separating pink and red. He saw these nuances more intensely than a non-color-blind person, like a sightless person who develops a hyper-sensitivity to touch and sound (we'll come back to this in a few pages).

The second moral concerns the process of scientific discovery. Dalton's strength wasn't an enormous power of reasoning, but an ability to feel that something wasn't right and not to stop searching until he'd found out what it was. Before it became a great scientific discovery, it was just a weird feeling. For tens of thousands of years, billions of color-blind people had experienced the same weird feeling without being able to put words to it.

And that's the third moral. We can peacefully coexist with people who don't see the same things as us, without ever becoming aware of the differences. The explanation is simple: we don't see inside their heads, and, literally, we don't see what they see.

When Dalton said that blue resembled pink, non-Daltonians didn't take him seriously. When Thurston said he saw in five dimensions, non-Thurstonians found it hard to believe.

There's an easy way to make sure that color-blind people aren't making fun of you (or, if you are color-blind, that non-color-blind people aren't making fun of you): the Ishihara test, which uses images formed of small circles of different colors. You see something different in the images, depending on whether you perceive that certain circles are or aren't the same color. In one of the plates in the test, color-blind people clearly see the number 21, whereas others clearly see 74.

There's no Ishihara test for the imagination. There's no direct way to verify that someone knows how to visualize the fifth dimension.

I don't know what Thurston really saw, but when I look at his mathematical work, I haven't the slightest doubt that he saw many things I don't. His style of writing gives the impression that he's just trying to share them with us. He'd love to show the things directly, in real life, but he knows that's impossible. So he writes math papers.

Seeing Is Finding Something Evident

In an interview with the *New York Times,* Thurston summed up things as follows: "People don't understand how I can visualize in four or five dimensions. Five-dimensional shapes are hard to visualize—but it doesn't mean you can't think about them. Thinking is really the same as seeing."

The last sentence, however, warrants some clarification. In the next chapter we'll discuss the nuance between a rapid and intuitive way of thinking, and a slow and reflective way. For a mathematician, "seeing" signifies thinking in a rapid and intuitive manner, directly, without need for reflection, as if the object really existed, as if it were right there in front of you.

Ease of access and immediacy count more than the visual nature of perception. Seeing is finding something evident. It's true etymologically ("evident" comes from the latin *videre,* which means "to see") and it's also true in everyday life: when you look at a block of ice, it's evident that it's cold, although you can't directly see its temperature.

An Ear for the World

Ben Underwood was born in California in 1992. When he was only two years old his mother saw a strange reflection in the back of one of his eyes. It was retinal cancer. When he was three, he had to undergo an operation to remove both his eyes. Sometimes people use the term "visually impaired" as a polite euphemism. In Ben Underwood's case, it's useless to hide behind a euphemism. He was blind.

When he was seven, Ben discovered he had a magic power: by clicking his tongue, he was able to see the world around him.

Real magic doesn't exist. Ben Underwood had simply learned to see by echolocation, like bats or dolphins. Clicks are sonar signals. Every object sends back a characteristic echo that tells you about its location, size, form, and what it's made of.

We already know that echoes can inform us about the space around us. To tell the difference between your bathroom and the inside of a cathedral, you just have to listen. And maybe you've already had the fascinating experience of waking up one morning and know-

ing, without even having opened your eyes, that it snowed during the night, because the texture of the silence has changed.

What we have a much harder time believing is that it's really possible, using only your ears, to reconstruct a reliable and detailed image of the world that surrounds us. It is, however, what Ben Underwood was able to do.

Without anyone explaining how to do it, without anyone even telling him it was possible, he developed a genuine faculty of "vision." Videos show him moving about freely, without using a cane or touching things. He does all the actions of daily life: going up and down stairs, opening doors, naming objects in front of him without picking them up, walking in the street, pointing at trees and their branches, riding a bike, roller-skating, weaving around cars, playing basketball.

Ben Underwood isn't the first sightless person to develop echolocation. The phenomenon has been known and documented for nearly three hundred years. But no one before him had brought it to this high a degree of perfection.

Ben Underwood shattered the limits of what was believed humanly possible. His abilities made him famous among the scientific community as well as the general public. As a teenager, he was invited on *The Oprah Winfrey Show* to share his story.

What might he have accomplished had he been able to pursue the development of his technique and continued to share his secrets? We'll never know. Ben Underwood died at age sixteen due to a recurrence of the cancer that had taken his sight.

The Laws of Mental Plasticity

Bill Thurston and Ben Underwood are striking geniuses. But what exactly is a genius? Is it a question of intelligence? Curiosity? Courage? Willpower?

I deeply admire both Thurston and Underwood, but it's not only to share my admiration that I'm telling their stories.

The real subject is our unfounded beliefs about the functioning of our brain. It's our illusion of being able to directly access the "real" world independently of the representation that our brain constructs of it. By ignoring the enormous range of possibilities available to us, we assign absurd limits to our intelligence. Ben Underwood's story is so hard to believe that we have the reflex to check on the internet that it's not some urban legend. That just gives a small idea of the fabulous experiences we deny ourselves.

The glaring omission of our culture and of our education is to teach us about our extraordinary mental plasticity, and that our destiny depends in great part on what we choose to do with it. The failure of math education is simply collateral damage of this omission. Those who don't have the chance to accidentally rediscover the actions that use this plasticity in the service of mathematics are condemned never to understand it.

Our general ignorance of the basic principles of mental plasticity is an enormous waste that reaches far beyond mathematics. Without pretending to know or understand everything, here are what seem to me the essential points:

1. *The power of our mental plasticity is profoundly shocking and almost supernatural.*

Stories like those of Bill Thurston and Ben Underwood are always hard to believe. They're amazing, but have a hint of sensationalism that's hard to shake off. You ask yourself what's the gimmick. But there isn't any gimmick. From a biological point of view, all of this is normal.

Our disbelief has a simple explanation, one that stems from the unconscious nature of the mechanisms at work. When someone tells us that Ben Underwood sees the world by analyzing the echoes of his clicks, we imagine him solving complex mathematical equations beyond the reach of a normal human being. That's both true and false. If you took a piece of paper and tried to solve the equations that govern the reflection of sound waves, you couldn't do it fast enough to see the world around you. It isn't believable that Ben Underwood could do the calculations in his head.

No human is capable of solving these equations in the way that school teaches us to solve equations, by applying a conscious and mechanical method. But the specificity of our mental plasticity is to give us an unconscious means of solving problems without ever stating them, by training our minds to recognize a multitude of subtle patterns that evade our consciousness.

No mathematician solves problems the way you're taught in school. It's biologically impossible to create truly innovative mathematics by following this method, just as it's biologically impossible to learn how to walk by solving Newton's equations.

How do you think you learned how to see, to walk, to talk, if it wasn't by a process that you weren't entirely aware of?

If you judged your own fundamental learning experiences using the same criteria that make you doubt the possibility of visualizing the fifth dimension or seeing the world by making clicks with your tongue, you would come to the same incredulous conclusion: rationally, learning to see, to walk, or to talk seems impossible. And yet you've managed to do it.

2. *The starting point is always insignificant.*

Try doing this: close your eyes and ask someone to place the palm of their hand directly in front of your face, then take it away and put it back without telling you while you make clicks with your tongue. You can hear the presence of the hand, just as you can hear the presence of a wall if you're within a few inches of it.

Starting with this primitive ability that you already have in a rudimentary state, you're free to develop your own ability for echolocation. It takes a kind of genius to dream it up yourself, but it's not the same thing once you know it's possible. Someone can even teach it to you, like Daniel Kish (sightless since early childhood, he, like Ben Underwood, invented his own technique of echolocation and teaches it today to young sightless people).

You have the same magic abilities as anyone else. It's just a question of will, patience, and openness to the world.

3. *Progress is slow and almost imperceptible.*

Mental plasticity is by nature a slow and invisible phenomenon whose progress is impossible to perceive in real time. It gradually transforms us, so gradually that, at first, we don't notice anything. At some point, however, we take notice, and it usually comes as a shock, precisely because we didn't see it coming. It happened unwittingly, in the background, without any effort on our part.

As for echolocation, it seems that you can get significant results by working an hour per day for two to three weeks. In the end, it's a bit like learning how to drive.

When you want to learn a new sport, a new language, or a new job, you go through a similar process. You have to throw yourself into it and accept that you'll be feeling about for a bit, thinking that you'll never be any good, until the moment you find, as if by magic, that you're getting the hang of it.

The Perfect Recipe for Discouragement

As a teenager, when my cousin Jerome bought a skateboard, I was shocked: "Why did he buy a skateboard when he doesn't know how to use it?" When he first got on it, he fell. Seen from outside, a person learning how to skateboard is just someone who spends all their time falling off it. Except that after a bit, as if by magic, Jerome was able to do it. Then it was more than just shocking. It had become unfair, scandalous, as if incompetence had been rewarded.

As long as you ignore the laws of mental plasticity, you underestimate others, and you underestimate yourself. The essence of mental plasticity is to transform audacity into competence.

The process is slow and invisible, and at first success seems unachievable: that's the biological reality of our learning mechanisms.

By an unfortunate coincidence, that's also the perfect recipe for discouragement. You need a lot of self-control and self-confidence to commit to a process that's confusing, slow, and uncertain.

That's why we so often limit ourselves to learning only what's officially possible to learn (things that have introductory or professional development courses), what you can learn by imitating others, or what comes naturally.

The rest, the secret and invisible apprenticeships, are said to be "gifts," "talents," "supernatural powers." No one tells us that we can learn to see in five dimensions, get our bearings through echolocation, or tell the sex of dogs and cats by looking at their heads, so we never even try.

We go so far as to fail to notice the "magic" powers that we develop without our knowing it: detecting insincerity in a smile or the sound of a voice, recognizing people we love through their particular scent, or knowing what they will say before they even say it. People who aren't good at math even forget that the video games they master in a matter of hours are cognitively a hundred times more difficult than high school math classes.

Reconnecting with your early childhood capacity for learning means to stop believing in these absurd stories of gifts and talent. It means to become once again capable of devoting ten or twenty hours to something that may or may not be impossible, without being distracted by the feeling of your own uselessness. It means to rediscover the world with an open mind, trying something just to see what happens, for fun, because you want to.

Ten or twenty hours doesn't seem like much. Seeing by echolocation sounds like a cool idea. If it only takes twenty hours, it seems worth the price. However, to spend twenty hours at something, you have to really want to do it.

Ten or twenty hours of real exploration, outside of our comfort zone, is enough to discover within ourselves unsuspected powers. But how many times lately have you spent ten or twenty hours at something entirely new?

The Great Hacking Project

When I was twenty-five, about one year before completing my PhD, I started looking at math as a pure activity of mental reprogramming, and I made the assumption that my mental plasticity had no limits.

Or, to speak more frankly, I chose at age twenty-five to throw myself into the deliberate and systematic project of hacking my cognitive abilities.

My basic technique hadn't changed: *lending an ear to the dissonance between my intuition and logic.* This technique remained my instrument for exploring the world, just like Ben Underwood's tongue clicks.

What changed at this moment of my life were my belief systems and the mindset that ensued. I stopped believing that our way of seeing and thinking about the world was a given fact, and that we each had a predefined amount of intelligence that we had to make do with. In place of that, I began to believe that we had the freedom to ceaselessly refashion our way of seeing and thinking, and to *construct our own intelligence* day after day.

In chapter 16 I'll talk about some of the increasingly extreme visualization exercises that allowed me to follow this path.

This change of approach had one initial practical consequence: I became a creative mathematician. I began to have ideas no one had had before, to see things no one had seen before, to prove theorems no one had yet proven—at first easy theorems, and later in my career theorems that had till then seemed far beyond my capabilities. Mathematical creativity has the reputation of being a great mystery that science can't explain. In my experience, however, it emerged as a natural phenomenon once I had adopted the correct psychological attitude.

But the greatest effect of this new approach was in my personal life. If I was able to hack my visual cortex and modify my way of perceiving space, if I was capable of changing even my way of understanding the notion of truth, what about all the rest? What about, for example, all that I'd believed were givens in my life, these "strengths" and these "weaknesses" that people spoke about and made up my so-called "personality"? What about my shyness, my mental blocks, my insecurities, and everything that was supposed to be holding me back? What about my social identity? How could these things be any less adaptable, less malleable, less freely reprogrammable than my perception of space and truth?

I remember with delight this beautiful day when, the very moment I stepped out into the street, I convinced myself that these things couldn't be fixed and determined, that they were necessarily open to reconfiguration, and that it was up to me to try.

Believing that you have a fixed personality, I thought to myself, is nothing but a superstition.

11

The Ball and the Bat

A ball and a bat cost a total of $1.10. The bat costs $1 more than the ball. How much does the ball cost?

This problem is taken from *Thinking, Fast and Slow,* the best-selling book by the psychologist Daniel Kahneman, winner of the 2002 Nobel Prize in Economics for his work on cognitive biases.

I encourage you to try the test with your friends. It works nearly all the time: most people answer that the ball costs 10¢. But that's not the right answer. If the ball cost 10¢, the bat would cost $1.10 (since it costs $1 more than the ball), and the ball and the bat together would cost $1.20.

If you explain why their answer is wrong, your friends will get it easily enough. But they won't necessarily know the right answer. They'll even find a lot of excuses: it's hard to do the calculations, they'd have to write down the equations but they don't have a pen, they can't be bothered. . . .

The correct answer is 5¢. If the ball costs 5¢, the bat costs $1.05, and together they cost $1.10.

The ball and bat problem plays a prominent role in Kahneman's book because it's the perfect illustration of his theory. According to him, we have two distinct cognitive systems, which he calls *System 1* and *System 2.*

System 1 allows you to give immediate and instinctive responses, without even trying. When someone asks you how much is 2 + 2, what year you were born, which weighs more, an elephant or a mouse, you don't even have to think. But it's also System 1 that makes you answer, incorrectly, that the ball costs 10¢.

System 2 is what you have to use when you're asked to calculate 47 × 83, or how many days have passed since your birth. You know how to get the answer, but you'd have to think. You probably need pencil and paper. One thing is certain: you don't really want to do it. Even if System 2 is more reliable and rigorous, you only use it when you have no other choice, because thinking hard, doing calculations, and logical reasoning are all tiresome.

Kahneman's theory can be summed up as follows:

1. Each time our System 1 gives us an answer, we're tempted to use it without calling on System 2, not even to verify that the answer is correct. Because System 2 uses a lot of mental energy and resources, we primarily rely on our instinct. Biologically, we've developed a preference for intellectual laziness.
2. In certain situations, our System 1 is systematically wrong. We all make the same mistakes, all the time, as if the wiring schematic in our brain was defective. These are the "cognitive biases" that Kahneman and his school have set out to study. We all want to say that the ball costs 10¢.

Kahneman's book became a best seller in part because it went beyond the simple theoretical observation and proposed a concrete method to avoid falling into the trap.

He has a simple recommendation: learn the list of cognitive biases presented in his book by heart, and each time you recognize one of the typical situations, fight your inclination and use your System 2 while trying to ignore your System 1.

I think I have a better way of doing it, which I'll explain.

"That's cheating!"

The first time I heard the story of the ball and the bat, it was from a friend who was studying cognitive science at Princeton. She had just read Kahneman's book and wanted to do the test with me.

Like most people, I gave an instinctive response. I listened to my System 1 without knowing it was called *System 1*. Without thinking, without doing any calculations, I gave the first answer that popped into my head: "5¢."

I felt that my answer annoyed my friend but I didn't immediately know why. She took the time to explain what was up. I was supposed to answer "10¢," or at least take a few seconds before answering "5¢." At any rate, there was no way I was supposed to answer "5¢" immediately, without taking the time to think about it. That was simply not allowed. A guy had even won the Nobel Prize for showing it was impossible.

Quickly enough, right before changing the topic of conversation, my friend did however come up with an explanation—a simple, pragmatic, and not entirely false one: "That's cheating! You're a mathematician!"

When I had my friends and colleagues take the test, I was sincerely surprised to find that so many of them answered "10¢," and even more surprised at their difficulty in finding the right answer after admitting that their initial response was wrong. The most incredible thing was that everyone spoke to me about "doing the calculations," as if it weren't visually evident that the right answer was "5¢."

I was like Dalton with his magic geranium, except that instead of being the one with the missing cones, I was the one seeing more colors than my friends. The other difference with Dalton, of course, is that the explanation doesn't have anything to do with genetics.

At the end of this chapter, I'll explain what I did to see the right answer—and how you can learn to do it as well.

A or B

This story about the ball and the bat really began to intrigue me, and I tried to understand what was stopping my friends from seeing the right answer when it was so obvious.

A bit like Dalton, I began my little inquiry. I believe I've found an explanation. After asking my friends about the price of the ball, I followed up with this question:

> Imagine that you have to make an important decision in your life. You have the choice between option A and option B. Your intuition tells you to choose A, but your reason tells you to choose B. What do you do?

I presented this question to more than a dozen of my nonmathematician friends, and almost all of them answered, without hesitation, that they'd follow their intuition and choose A. Only one person chose B. Another hesitated for a long time, without ever giving a clear response.

Nothing guarantees that you'd get such a high percentage of people choosing A if you tried this experiment on your own. My protocol suffers from what is called *selection bias,* in that my friends aren't necessarily representative of the general population, and it may very well be that people who listen to their intuition have greater chances of becoming my friends.

The exact proportion of A and B didn't really interest me. What I wanted to know was whether someone would come up with the same response I would have given myself. No one did.

My hypothesis is that my unusual response to the question is the key that allowed me to become good at math and, along the way, to reeducate many of my cognitive biases.

An Unreasonable Assumption

Kahneman says that thousands of American students took the ball and bat test, and that "the results are shocking." At the lower-tier universities, the error rate was over 80 percent. Even students at Harvard, MIT, and Princeton gave the wrong answer more than 50 percent of the time.

Kahneman's book is fascinating, but I'm confused whenever I see him opposing "the right answer" and "the intuitive answer," as if there were only one intuitive response possible, and it was necessarily false. For example, he writes: "It is safe to assume that the intuitive answer also came to the mind of those who ended up with the correct number—they somehow managed to resist the intuition."

In other words, Kahneman finds it safe to assume that I shouldn't exist. My opinion, understandably, is that this isn't a reasonable assumption.

But beyond the relatively minor question of my own existence, this anecdote reveals a major disconnect between Kahneman's theory and what all mathematicians know deep in their bones. It's up to you to decide who's in the best position to give you advice on mental calculations.

Kahneman finds it shocking that 50 percent of the students at Harvard, MIT, and Princeton blindly relied on a manifestly false intuition, and I'm as shocked as he is.

But I'm equally shocked by something that Kahneman apparently finds completely normal: how is it that 50 percent of the students at Harvard, MIT, and Princeton managed to get accepted despite having such faulty intuitions?

Having studied and taught at highly competitive universities, I know that students who can directly "see" the correct answer in their head have an enormous competitive advantage. I don't understand how the others can even compete. I imagine that they compensate by intensive cramming, something I'm completely incapable of and the very thought of which gives me a headache.

Kahneman's advice consists of identifying the situations where we should "resist" our intuition and submit ourselves to System 2. It's strange advice coming from someone who's spent his life documenting our aversion to effort, our preference for instinctive and immediate responses, our immoderate love for System 1—and our hatred of System 2.

This idea that we should resist our bad instincts and entirely submit to a robotic mode of thought was once the prevalent paradigm in education. Kahneman is well enough placed to know why it can't work.

Another point bothers me. It's true that we should be wary of our System 1. But what are we to make of our System 2? Personally, I stopped trusting mine after ninth grade, when I found out I wasn't able to string together three lines of calculations without making a mistake.

But the most troubling aspect is that Kahneman reasons as if our intuition were hardwired, with no possibility for us to reconfigure or reprogram it. Had he lived in the ancient Roman era, he would almost certainly have said that it was impossible to represent mentally the result of the operation "1,000,000,000 – 1," because the number greatly exceeded the capacities of human intuition.

System 3

When I need to make an important decision in my life, if my intuition tells me to choose option A and my reason tells me to choose option B, I tell myself there's something going on and I'm not ready to make the decision.

That's the moment to resort to what I call *System 3*.

System 3 is an assortment of introspection and meditation techniques aimed at establishing a dialogue between intuition and rationality. You use it each time you try to recall your dreams, to put words to the fleeting impression that left a strange taste in your mouth, to sort out your most confused and contradictory ideas.

When I was eighteen and I discovered that the stupid images in my head had a tendency to correct themselves once I made the effort to describe and name them, when I got into the habit of *lending an ear to the dissonance between my intuition and logic,* I put System 3 at the center of my strategy for learning math. The results exceeded my wildest expectations.

We all know System 3 and we all use it, at least from time to time. My mathematical journey taught me that a voluntary and radical use of System 3 is not only possible, it augments our intuitive capacities well beyond the supposed limits of human cognition.

Through the years, the systematic search for a better alignment between my intuition and logic has become my way of understanding the world, others, and even myself.

In practical terms, here's what that means. When my intuition tells me A and rationality tells me B, I put myself in the position of a referee. I force myself to translate my intuition into words, to tell it like a simple and intelligible story. Vice versa, I try to picture what logical reasoning is actually expressing, to experience it in my body, to hear what it's trying to say. I ask myself if I really believe it. I fumble about. It takes time but it's not a real effort. It's more like a meditation on running water, something going on in the background that might stop and start, then all of a sudden become clear days, months, or even years later.

The goal is to understand where things are going wrong. Are my intuition and logic even speaking the same language? Are they even talking about the same things?

My intuition is never perfect. It's often relevant, but sometimes it's just rubbish. The good news is that it's generally fixable. As for logic, that's never wrong. At least officially. Except that it doesn't necessarily say what I think it's saying.

In the end, it's almost always my intuition that wins. When I force it to listen to what logic is saying, it takes that into account and adjusts its position. Logic is something inert, like a pebble. My intuition is organic, it is living and growing.

It's obviously stupid to call this approach *System 3*. It should simply be called *thinking* or *reflecting*. But the meaning of these words has been hijacked by a tradition that wants to make us believe that we should think contrary to our intuition. We're told that our intuition is the mortal enemy of reason, that any dialogue between the two is impossible, and thinking means you have to submit blindly to System 2.

I'm personally incapable of thinking against my intuition and I have serious doubts as to the sincerity of people who claim they can.

In chapter 3 I said that your intuition was your most powerful intellectual resource. However, at the risk of spoiling your dreams, I must be honest with you: your intuition isn't a magical elixir, or your lucky star, or the hand of God on your shoulder. It's much more trivial than that. It's the tangible manifestation of a reality that is invisible but perfectly concrete and material: the entanglement of synaptic connections between your neurons that your brain continuously constructs and reorganizes, as it has done since you were in the womb.

Your brain contains as many neurons as there are stars in the Milky Way. Each of these neurons is, on average, tied to thousands of other neurons. This fabric of hundreds of trillions of interconnections is the network of your mental associations. Its structure is your way of giving meaning to the raw information continually flooding into your brain. This is, literally, your vision of the world. All that you have seen, heard, felt, imagined, or desired, all of your experience, all that you know, all that you remember, is encoded in this web. When your intuition speaks, that's where it's speaking from.

Your intuition will always be more powerful and better informed than the most sophisticated of language-based reasonings. For all that, it's not infallible. If your intuition tells you the ball costs 10¢, it's plainly wrong.

My intuition isn't any less fallible than yours. It's always getting things wrong. I have, however, learned never to be ashamed of it. I don't disdain my mistakes, I don't push them aside, because I don't think that they betray my intellectual inferiority or some cognitive biases hardwired in my brain. On the contrary. Nothing's more exciting than a big glaring error: it's always a sign that I'm not looking at things in the right way, and that it's possible to see them more clearly. When I'm able to put my finger on an error in my intuition, I know it's good news, because it means that my mental representations are already in the process of reconfiguring themselves.

My intuition has the mental age of a two-year-old—it has no inhibitions and always wants to learn. If you stop mistreating your own, you'll see that it's exactly like mine, only asking to be allowed to grow.

The Price of a Ball

Because I have terrible handwriting, and because I'm easily distracted, I have a tendency to make mistakes in calculations.

I discovered in ninth grade that the only way to get around that was to verify after every three lines that what I was writing still made sense and that I really believed it. In other words, I learned how to use my System 1 to supervise the work of my System 2. From this time on, I was incapable of manipulating mathematical objects that I had no intuition for.

At what moment did I stop primarily visualizing numbers through their written form? I don't recall. But it undoubtedly goes back to the same period. Decimal writing of numbers is useful for written calculations but it is certainly less practical when you want to form an intuitive idea about the validity of those calculations. This is where System 1 has an edge: it isn't bound by the constraints of language and writing.

Depending on the context, I have many different ways of visualizing numbers. I have, for example, a tendency to visualize price in terms of length. When my friend told me that the ball and bat together cost $1.10, I immediately translated her words into a mental image that looked something like this:

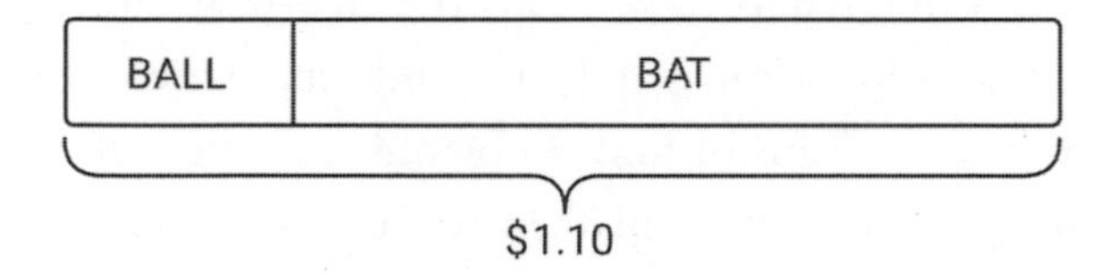

When she told me that the bat cost $1 more than the ball, here's how I saw it:

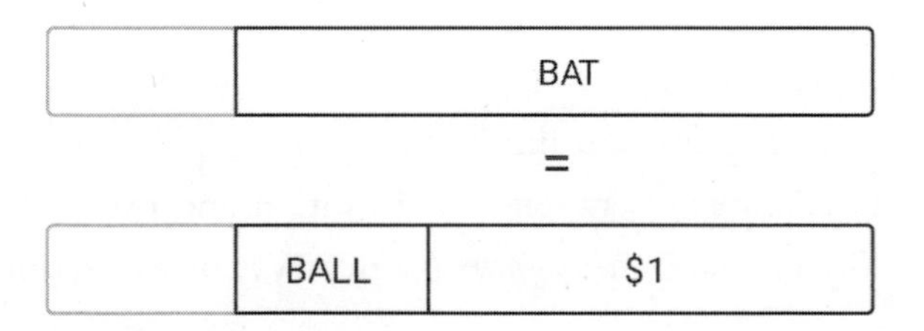

Then the two images came together in my head and morphed into something like this:

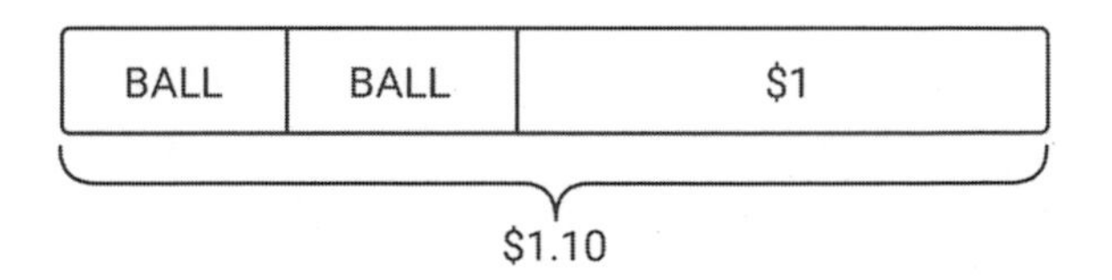

If this is how you visualize the problem, it doesn't take a genius to figure that a ball costs 5¢.

A mental image is neither good nor bad in and of itself. Its value lies in what it allows you to understand. There are countless ways to visualize the problem and I don't pretend that mine is better. My numeric intuition isn't all that remarkable. I wouldn't be able to do the calculation if the ball and bat together cost $2,734.18 and the bat cost $967.37 more than the ball.

I have these pictures in my head because, in my life, I've made a lot of calculations errors. Instead of concluding that I was terrible at math, I simply looked for simpler ways to see things, to grasp what I was writing.

In time, with this approach, I constructed a great variety of mental images that help me today to better understand the world.

If you want to learn to find it obvious that the ball costs 5¢, I recommend proceeding like a mathematician would when faced with a new and incomprehensible idea. Rather than learning my pictures by heart, train yourself to construct pictures that work for you. The most important messages, the ones you should always bear in mind, are these:

1. You can reprogram your intuition.
2. Any misalignment between your intuition and reason is an opportunity to *create within yourself* a new way of seeing things.
3. Don't expect it all to come at once, in real time. Developing mental images means reorganizing the connections between your neurons. This process is organic and has its own pace.
4. Don't force it. Simply start from what you already understand, what you can already see, what you find easy, and just play with it. Try to intuitively interpret the calculations you would have written down. If it helps, scribble on a piece of paper.

5. With time and practice, this activity will strengthen your intuitive capacities. It may not seem like you're making progress, until the day the right answer suddenly seems obvious.

You'll need a number of training sessions. Exactly how many, I don't know. It's not worth tiring yourself out—better to split it up into short five-minute sessions, and think about it in the shower or while on a walk. Above all, take your time. It's good to think about it only once a week or once a month. Most important, keep at it and don't let it drop. It will come eventually.

Solving a problem is only ever a pretext. The important thing is that you have the power to reeducate your intuition, to gain confidence in your body and thoughts.

Nothing about this should surprise you. Solving the problem of the ball and the bat is like standing up on a surfboard. Kahneman says that the first time you stand up on a surfboard, you'll fall in the water, and concludes that humans are born with a defective sense of balance and that getting up on a surfboard can never become intuitive. His advice is to get out of the water and learn the laws of physics by heart. My advice is to get back up on the board.

Electrical, Mechanical, Organic Thought

A central idea of this book is that our culture conveys false beliefs about how our brain functions, and that these false beliefs keep people away from the simple actions that would allow them to become good at math.

When you say to people that certain truths are by nature counterintuitive, you tell them that they can never really understand. It's a way of discouraging them. Nothing is counterintuitive by nature: something is only ever counterintuitive temporarily, until you've found means to make it intuitive.

Understanding something is making it intuitive for yourself. Explaining something to others is proposing simple ways of making it intuitive.

None of this takes away from the value of Kahneman's work. The cognitive biases that he documents are striking human realities of great social importance. We all have biases, even if they aren't hard-coded and vary from one person to another, and certain biases happen to be more widespread and problematic than others.

The distinction Kahneman makes between System 1 and System 2 has the merit of being simple. In a sense, it picks up the classic opposition between left brain and right brain, but in a modern version, without the anatomical nonsense. It's just a basic model, but it's appealing, and helps us become aware of our different ways of mobilizing our mental resources.

We'll end this chapter with a summary of the characteristic principles of System 3, the big oversight in Kahneman's theory. We'll talk again in chapter 19 about the physical structure of our cortex and how it functions, which will allow for a biological interpretation of System 3 and its efficacy. At any rate, System 3 is a good model for the real nature of mathematical work.

System 1 is our intuitive capacity. We all like to describe it using electrical metaphors: with our intuition, we say that we think with *the speed of lightning*. It's not entirely false. Our brain isn't properly speaking an electrical circuit, but the signal that's transmitted along the neurons is electrical in nature.

System 2 is our capacity for rigorous reasoning. We imagine it in mechanical terms, with gears or something of the sort. That doesn't correspond to any biological reality. What we are biologically capable of is to pretend we're robots and mechanically apply a preset series of instructions. With the right set of instructions, we can make logical conclusions and valid calculations. But it's so disagreeable and against

Thinking fast, slow, and super slow

	System 1	*System 2*	*System 3*
Name	Intuition	Rationality	Thought?
Verb	See	Follow the rules	Reflect? Meditate?
Adjective	Instinctive	Procedural	Introspective
Output	Mental image	Calculated value	Updating System 1
Speed	Fast	Slow	Super slow
Time scale	Immediate	Seconds or minutes	Minutes, hours, days, months, years
Metaphor	Electrical	Mechanical	Organic
Benefits	Speed, facility, sincerity	Accuracy	Strength, tranquility, self-confidence
Limitations	Imprecise and incoherent	Not human	Asynchronous

our nature that we usually give up after a few seconds, or at most a few minutes. In the end we're rather sorry robots: we make too many mistakes and we can't go the distance.

System 3 is so entirely ignored by our culture that I can't find the right word to characterize it. As I said above, I would like to say that System 3 simply corresponds to our capacity for *thinking*. But the verb *to think* doesn't mean much since it's been used as an injunction to submit to System 2.

The activity of System 3 is a special kind of *meditation*, but this word is also much too vague. Not all meditation is an activity of System 3. System 3 specifically aims at establishing a dialogue between Systems 1 and 2, in order to understand their misalignments and resolve them. Rather than a free meditation, it's one constrained by the principle of noncontradiction. Its ultimate goal is to revise and update System 1 while taking into account the results of System 2.

It's also necessary to distinguish System 3 from the capacity of our System 1 to revise itself without any deliberate action on our part. Our mental plasticity results from the constant reconfiguration of our synaptic network: our mental circuits evolve in response to our experiences. You can imagine neurons as minuscule plants that grow and sink their roots deeper and deeper.

Every time we practice a given activity we habituate our System 1 to the specifics of that activity. When we try to stand up on a surfboard, we habituate our System 1 to the hard realities of Newtonian physics and construct our surfing instincts. With System 3, we habituate our System 1 to the hard realities of logical consistency and construct our instincts for truth.

The great misunderstanding of math teaching stems from the fact that all the tangible manifestations of mathematics—its confusing language, its incomprehensible notation, its bizarre and rigid reasoning—seem to tie it to System 2.

Most people take that at face value. They become discouraged in a few minutes, or throw themselves into a masochistic effort that has zero chance of succeeding.

But a few people choose to rely on their System 3. They're not aware that they're doing anything special. Math just feels easy to them. It doesn't even feel like work. They're just seeing pictures in their heads, and spending a couple of minutes a day looking at these pictures and asking themselves naïve questions.

To them, it all looks completely normal. It certainly doesn't feel like they have a gift.

12

There Are No Tricks

It's an ordinary day in the United States in the early 1950s. An ordinary family is on an ordinary road. The father is driving; the two kids are in the backseat. To stop them squabbling, the father asks them some puzzles:

What is 1 + 2 + 3 + . . . + 100?

The younger boy is five years old. In a few seconds he answers "5,000." The father tells him that's almost it. The young boy thinks for a few seconds more and finally gives the right answer: "5,050."

The five-year-old boy is Bill Thurston. The story makes you smile, especially if you know the famous tale about Carl Friedrich Gauss (1777–1855), "the prince of mathematics." Even if this old story is nothing but a legend, it's very well known, and Thurston's father undoubtedly had heard about it.

Gauss was one of the greatest mathematicians in history, someone you could without hesitation place alongside Thales, Pythagoras, Euclid, Archimedes, al-Khwarizmi, Descartes, Euler, Newton, Leibniz, Riemann, Cantor, Poincaré, von Neumann, Grothendieck, and a few others. He was so spectacularly brilliant and creative that his contemporaries refused to believe that his intelligence came from a biologically normal human brain. He was in a way the Albert Einstein of his time.

And, in fact, it ended exactly as with Einstein. When Gauss died, someone thought it wise to take his brain in the hope of uncovering its secrets. Two centuries later, Gauss's brain is still preciously conserved in a jar, somewhere in the collections at the University of Göttingen. No one has of yet found anything particularly interesting to say about it.

The legend has it that, at the prime age of seven, the young Gauss scared the hell out of his instructor. The latter had asked the class to calculate the sum of whole numbers from 1 to 100, in the belief that he'd be giving himself a good quarter of an hour's peace. He hadn't counted on one of the kids finding the answer in seconds.

I was seventeen when our senior high school math teacher told us this story, which had a big impression on us. We couldn't figure out how Gauss was able to calculate so quickly. Faced with such genius, we all felt ourselves rather pathetic.

The explanation that our teacher gave us was that there was a "trick." You want to calculate the whole numbers from 1 to 100—that is, to add them up:

$$1 + 2 + 3 + 4 + \ldots + 97 + 98 + 99 + 100$$

The trick consists of doubling this sum by counting each whole number twice and placing the two sums on two lines, in the following manner:

$$\begin{array}{rrrrrrrrrr} & 1\,+ & 2\,+ & 3\,+ & 4\,+ & \ldots\,+ & 97\,+ & 98\,+ & 99\,+ & 100 \\ +\; & 100\,+ & 99\,+ & 98\,+ & 97\,+ & \ldots\,+ & 4\,+ & 3\,+ & 2\,+ & 1 \end{array}$$

What a strange idea! Why put down each number twice? Why place them one above the other in this bizarre fashion? Maybe it is strange, but you have the right to do it. At any rate, each number from 1 to 100 does appear twice. The value of the big sum is thus double the number that we're looking for.

Now instead of looking at the lines, look at the columns. There are 100 columns and in each there are two numbers whose sum is always 101. It seems like magic, but it's true. The big sum thus equals 100 × 101, or 10,100. The number we're looking for is half of that, or 5,050.

Don't be ashamed of having to read over this reasoning a number of times before you find it convincing. As with all mathematical reasoning, there's something bizarre and intimidating about it. At first you have to decipher it line by line, which takes a lot of time and effort.

The steps of the reasoning, however, are simple enough, and should allow you to reach these three conclusions:

1. It's a valid proof of the fact that the sum of whole numbers from 1 to 100 equals 5,050.
2. It's believable that someone quick in mental calculations could do this reasoning in their head in a few seconds.
3. But how on Earth could such a crazy idea arise in the head of the seven-year-old Gauss?

At any rate, those were the conclusions I made myself when I was seventeen. I figured out that math wasn't for me because it was meant for those other people, those *geniuses,* whose brains worked differently than mine and could come up with such incredible ideas.

My teacher was an excellent instructor, and I'm grateful for everything he taught me. But that day, by telling us there was a "trick," he sent us the wrong message.

There are no tricks. There never were any and there never will be. Believing in the existence of tricks is as toxic as believing in the existence of truths that are counterintuitive by nature. These are the two central superstitions of the System 2 dogma, this belief that our intuition isn't worth a dime and that we have to mechanically apply methods that we don't fully understand.

Of course it can happen that things work without our understanding why. It happens often enough. But it's always a temporary situation that's just waiting for an explanation.

Believing that tricks exist is to accept the idea that there are things you'll never understand and that you have to learn by heart. It's to confuse the line-by-line verification of a proof with its intuitive understanding. It's to enter into a submissive relationship to System 2. It's to accept a division of roles that is deeply unfair and humiliating for you: the great geniuses find the tricks, while you're only good for checking that it all adds up.

Frankly, I couldn't care less about verifying that the sum of whole numbers from 1 to 100 really equals 5,050. What I want to know, what we all want to know, is how to think like Gauss and Thurston.

The Language Trap

To understand what's hidden behind these math "tricks," the simplest thing is to follow a recipe for banana bread.

Ingredients:

1 1/2 cups (195 grams) all-purpose flour
1 teaspoon baking soda
1/4 teaspoon fine sea salt
3/4 teaspoon ground cinnamon
3 medium bananas
8 tablespoons (115 grams or 1 stick) unsalted butter, melted and cooled
3/4 cup (150 grams) packed light brown sugar
2 large eggs, lightly beaten
1 teaspoon vanilla extract

Directions:

1. In a large bowl, mash the bananas with a fork.
2. Combine the flour, baking soda, and salt.
3. Cream together the eggs and sugar.
4. Stir in the mashed bananas, vanilla, butter, and cinnamon.
5. Stir in the flour mixture, a third at a time, until just combined.
6. Pour the batter into a 9-by-5-inch loaf pan. Bake for about 1 hour at 350 degrees F.

Visualize the different steps of this recipe:

—You start by buying the bananas. They're in your hand. You go to the register to pay for them. Can you picture them?
—You're at step 1. The bananas are in the bowl. You have a fork in your hand and you're getting ready to mash them. Can you still picture them?

Between these two steps, you've switched to a different mental image. Just before mashing the bananas, you've mentally peeled them. Behind the so-called "tricks," there's generally an operation of this kind: a change of mental image that's done in a flash, for "obvious" reasons that might not be so obvious to anyone else.

When you're familiar with bananas, it's obvious that you have to peel them before you mash them. But if you've never seen a banana before, it's not so obvious. Recipes never capture all the steps you need to take. There are always some missing details, the famous "tricks." It's why so many people prefer watching a cooking video to reading a recipe.

You've been acquainted with bananas since childhood. You could even say that you've developed some kind of spiritual intimacy with them. You know a lot of things about them that you've never told anyone. You know by heart the string that runs along the flesh even if you don't know its name. You've never done anything with this string, but it's always struck you by its appearance and properties. You also know, without ever having dared say it, that nothing on Earth squishes in such a soft and satisfying manner as the flesh of a banana. The word *banana* doesn't evoke just one mental image but a multitude of possible mental images. Instantly, without even trying, and without anyone telling you how to do it, you always pick the right image. Mashing bananas without peeling them is so idiotic that you find it funny. It's the stupid kind of thing that only a robot would do.

When Gauss or Thurston wanted to add up the whole numbers from 1 to 100, they picked the right way to visualize these numbers, the way that made the calculation easier. They found it instantly, without any effort, and with no one to tell them how. They knew how to mobilize their familiarity with numbers in the same way that you know how to mobilize your familiarity with bananas. It's exactly the same kind of intelligence.

In mathematics, the sudden occurrence of a miracle or an idea that seems to come out of nowhere is always the signal that you're missing an image. Your way of looking at things isn't the right one. Something is missing. There exists a better way, simpler, clearer, deeper, that you don't know yet and that, perhaps, no one yet knows. Looking for and finding the right way of seeing things is the driving force of mathematics. It's the main source of pleasure you can take from it.

Each time someone talks to you about "tricks," they're telling you to stop thinking at precisely the moment when it starts to get interesting.

The irony of all this is that while you were just getting familiar with bananas, at that faraway time of your childhood, you were also getting familiar with numbers. If you hadn't developed the right level of intimacy with numbers, you would never have been able to learn how to count.

Unfortunately, you've since lost this intimate relationship to numbers. Following your early childhood, you fell into what I call the *language trap,* which is what stops you from "seeing" the sum of whole numbers from 1 to 100 like Gauss or Thurston.

The language trap is the belief that naming things is enough to make them exist, and we can dispense with the effort of really imagining them.

This belief is typical of the ideology of System 2. We're told that we should think with words and that yearning to move beyond words is a pipe dream. This shortcut is problematic, if not an outright lie. Naming things certainly allows us to evoke them, but not to make them present in our mind with the intensity and clarity that allow for creative thinking.

"Don't think of a pink elephant." This is considered a linguistical paradox, since the sentence itself forces us to think of a pink elephant. Except that this passive, reluctant way of thinking about pink elephants isn't one that will allow you to get to know and really understand them. Try to imagine a life-sized pink elephant standing before you. Take the time to look at it and study it closely. This intentional image will be incredibly more profound, more absorbing, more precise than the fuzzy image formed in your mind at the beginning of this paragraph. When you give free rein to your imagination, it is nearly without limits.

It's this effort of the imagination that allows you to get out of the language trap and solve mathematical problems. This activity is at the heart of System 3. It implies deliberately trying to see, without reserve or half measures, with a full physical commitment.

When you read "the sum of whole numbers from 1 to 100," if you content yourself with the fuzzy image that forms in your head, you won't really see anything.

Instead of letting yourself be lulled by words, force yourself to think that the sum is physically present in front of you. Force yourself to imagine the whole numbers from 1 to 100 in physical form, made manifest in the real world, carefully lined up in front of you. If you manage to see them and you take the time to carefully examine the scene, you'll find a way to calculate their sum.

To give you a chance to find it for yourself, I recommend you take a short break before continuing.

The Big Picture

In "On Proof and Progress in Mathematics," the text cited in chapter 6, Thurston gives some surprising advice—that I've never read anywhere else—about the size of mathematical objects.

When we imagine them in our heads, we can choose to see them as "little objects in our hands," or as "bigger human-sized structures," or as "spatial structures that encompass us and that we move around in." From a logical standpoint, it shouldn't make any difference. Thurston, however, says size is quite important: "We tend to think more effectively with spatial imagery on a larger scale: it's as if our brains take larger things more seriously and can devote more resources to them."

What if people who are convinced that they don't have any geometric intuition simply make the mistake of imagining figures that are too small, so that it's impossible to see anything?

At any rate, Thurston's remark applies very well to elephants: imagine a tiny elephant that you can hold in your hand, and now imagine a life-sized elephant, who doesn't look happy, and whose attention you don't want to attract: it mobilizes your cognitive resources in an entirely different manner.

The language trap is the extreme version of the phenomenon

described by Thurston. An expression like "the sum of whole numbers from 1 to 100" is a convenient way to designate a very precise mathematical object. It allows you to speak of it, but it's also a way of getting rid of it, to put it at some distance, so that it doesn't bother you any longer.

You think you see the sum, but you don't really. You can't feel its looming presence. You don't take it seriously.

This sum can also be written as 5,050. The big advantage of decimal writing is that it's compact. It's discrete, practical, easy to say and easy to write. This mental representation has its weakness in its strength: the number is put at such a distance that it becomes minuscule, almost invisible.

A mathematical equation always states that two writings that are different in appearance designate in reality one and the same object. If you let yourself be lulled to sleep by language, if you confuse words with the objects they designate, you don't give yourself any chance to "see" mathematical equations.

The only way to get there is to go beyond words. Replacing "the sum of whole numbers from 1 to 100" with "1 + 2 + 3 + . . . + 98 + 99 + 100" is a good start. You might have the impression of seeing the sum in a more tangible and concrete way. But that's only ever an illusion. In reality, you'll be missing most of the numbers, those hidden by the ellipses. Mathematical symbols are like words, they belong to language. You also need to get beyond them.

To see the sum in its entirety, without shortcuts or abbreviations, to take it seriously and give it the place it deserves, you need to imagine it as physically present, life-sized, right in front of you.

Before imagining the sum, let's start with a single number, for example, 3. Imagining the number 3 in the physical world is easy enough: you just have to imagine three objects, like in primary school when the teacher asks students to picture three oranges *in their head.*

This childlike relationship with numbers, this need for a *bodily* interaction with abstract things, is the right state of mind to do mathematics. By seeing three oranges in place of the number 3, you begin freeing yourself from the language trap. You stop confusing the writing of a number with its value.

Mathematicians have a saying that a whole number always counts for something. However, to imagine the sum of whole numbers from 1 to 100, I'd advise not using oranges: you'd find yourself with a lot of oranges—they will spill all over the place and it will be a mess.

Personally, I find it much easier to imagine the scene using cubes. I'm able to visualize each number as a pile of cubes and to arrange these piles side by side, from 1 to 100.

It's hard to draw exactly what I see in my head. My mental image isn't entirely distinct, and the pile would be too big to put on the page. I'm only able, therefore, to draw an approximation. Seen head on, it would look something like this:

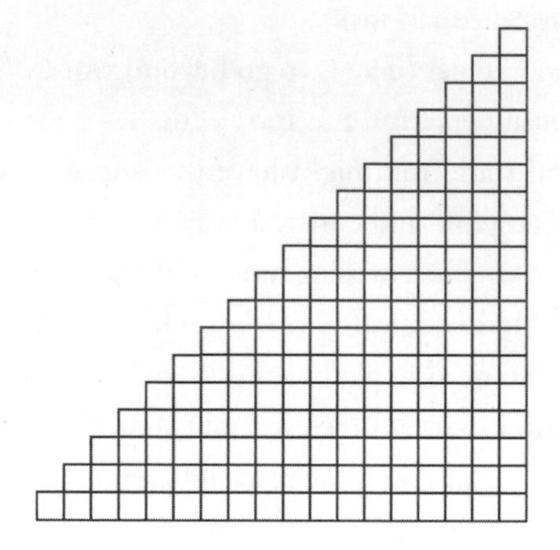

The drawing is wrong, but that doesn't matter. What matters is to know in what way it is wrong. In this case, there are some cubes missing. Instead of having eighteen cubes across and eighteen cubes high, as in my drawing, the pile should have one hundred cubes across and one hundred cubes high. You need to keep that in mind. Despite that, this drawing seems to me to be a good way to share my mental image (if I drew all the cubes, they would be too tiny).

Voilà! We're done. Wasn't that easy?

Mathematicians tend to think that a proof is finished once they feel that the right image is formed in their heads, like when chess players stop the game before checkmate once they see one player has a winning position.

But let's take the time to finish the game, as the coming checkmate may not yet be that obvious to you.

If you have this image in your head, it's difficult not to see a triangle. The number you're looking for, that is, the total number of cubes, is the area of the triangle. There is a simple primary school equation to calculate the area. Here are two ways to finish the game, depending on whether you know the equation.

1. *You know the equation.* To calculate the area of a triangle, you multiply the base by the height and divide by 2. Here the base is 100 and the height is 100. Multiplying them gives us 10,000 and dividing that by 2 gives us 5,000.

We're almost there. We just made the same error Thurston did when he was five. That's a good sign; we're surely on the right track.

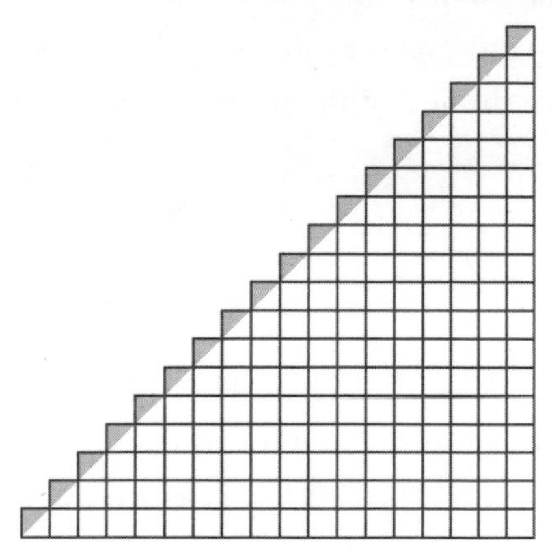

The error was forgetting about the half cubes that are above the diagonal and aren't counted in the area of the triangle. We've forgotten 100 half cubes, so we have to add 50, which makes 5,050.

2. *You don't know the equation.* No worries, you're going to reinvent it. By looking carefully, you can see that a triangle is a half of a rectangle. If you take your initial triangular pile (in white) and a copy of this triangular pile (in gray), and then turn the copy and put it on top of the initial pile, you come up with something like this:

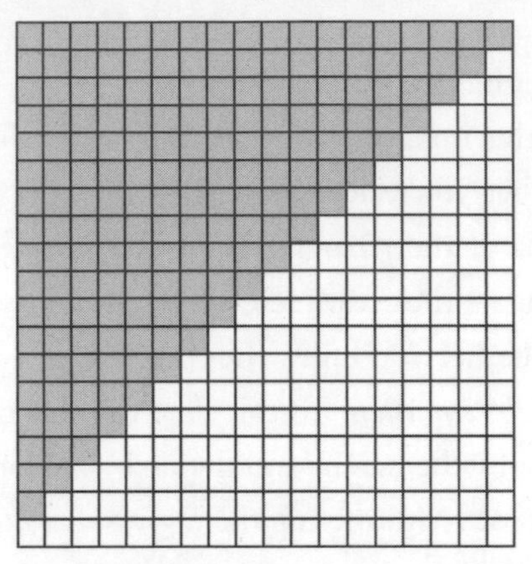

You thus get a rectangle 100 cubes across and 101 cubes high, therefore formed of 100 x 101 = 10,100 cubes. There are thus 5,050 cubes in each triangle. The famous "trick" that consists of taking the two sums and putting them one over the other is nothing more than that, a way of breaking down the area of the rectangle into the two triangles:

$$\begin{array}{rrrrrrrrr} 1 + & 2 + & 3 + & 4 + \ldots + & 97 + & 98 + & 99 + & 100 \\ + 100 + & 99 + & 98 + & 97 + \ldots + & 4 + & 3 + & 2 + & 1 \end{array}$$

Probabilistic Kung Fu

The ball and the bat, the sum of whole numbers from 1 to 100: I love these problems because they are elementary, and yet they fully illustrate the gap that separates official mathematics, the prisoner of language, and secret mathematics, what you do inside your head.

In both cases, a simple visualization effort suffices to make something easy that 99 percent of people don't find easy at all.

It's not always so simple. Visualization isn't always enough, and the issue isn't to deprive yourself of mechanical deductive reasoning. In order to understand mathematics, you have to train yourself to bring together imagination and language, intuition and logic, reverie and calculation, seeing both the big picture and the details.

Nor do I want to give the impression that math problems are always numerical and that all intuition is geometric in nature.

Mathematical objects are very diverse in nature, and their intuitive comprehension mobilizes different mental resources. The table enumerates several of the main mathematical domains. It's incomplete and oversimplified, but it gives you a general idea.

These domains each have their own vocabulary and intuitions. It's like they correspond to different ways of using our bodies, different

The main areas of mathematics

Domain	*Objects studied*
Arithmetic	Whole numbers
Geometry	Space and shapes
Topology	Spaces and shapes that can be stretched and twisted
Group theory	Symmetries and transformations
Algebra	Abstract structures
Analysis	Limits, "infinitely small things"
Probability theory	Chance, randomness
Logic	Proofs (seen as mathematical objects)
Algorithmics	Procedures and computations (seen as mathematical objects)
Dynamical systems	Things that evolve with time
Combinatorics	Ways to count objects

regions in our brains, different ways of focusing our attention. They might give the impression that they're talking about different things, but in fact they're simply bringing different points of view to the same reality, the mathematical reality. When you get to experience it, the unity of mathematics is absolutely fascinating. It can even be overwhelming.

Very often, mathematical discoveries are merely bridges between two intuitions of different kinds.

At a very elementary level, this is what we've just done above: a geometry formula (that gives the area of a triangle or rectangle) allowed us to solve an arithmetic problem (the sum of whole numbers from 1 to 100).

Let's conclude this chapter with another, even more striking example.

If you have problems visualizing whole numbers from 1 to 100 life-sized before you, there's an easier way. Instead of tiring yourself out by dealing with all the numbers from 1 to 100, why not take only one of them, at random? When you randomly choose a number between 1 and 100, what is its value *on average?*

If that seems abstract to you, here's a concrete way of imagining it. You're on a game show. In a bag, there are one hundred checks: a check for $1, a check for $2, and so on up to a check for $100. You can pick only one check, eyes closed. On average, how much do you expect to win?

I'll repeat the question: *when you choose a number at random from 1 to 100, what is its value on average?*

Most people will answer "50" without thinking. It just seems obvious to them. But if the average of whole numbers from 1 to 100 is 50, then their sum must be 5,000: the sum of 100 numbers is 100 times their average. That's also obvious to most people.

So what is it that keeps people from answering without thinking, like the five-year-old Thurston did, that the sum of whole numbers from 1 to 100 is 5,000?

(If your intuition does tell you that the average is 50, it's slightly wrong. Don't worry, it's exactly the same mistake Thurston made. The average is actually 50.5. At this stage, an error of 1 percent shouldn't spoil your fun.)

What just happened? How is it even possible? How come the difficulty of the problem vanished all of a sudden, like it was vaporized? If this seems absurd to you, it's because you're underestimating the power of probabilistic thinking. Assembling 5,050 cubes in your head is a demanding task, it almost feels like you need a forklift. In contrast, the probabilistic approach is a sort of kung fu that focuses your attention on a single number while your subconscious does all the heavy lifting.

In fact, you already knew how to calculate the sum of whole numbers from 1 to 100, even if you were not aware that you did.

The notion of *average* is a purely human invention, an abstract mathematical concept that you were taught and that you've assimilated in the deepest part of yourself, just like decimal writing. You've learned to "see" averages, that is, to calculate them without even thinking or having to write them down. If you want to validate your intuition and transform it into rigorous thinking, if you really want to understand why the average is 50.5 rather than 50, you need to listen to yourself, to your unconscious processes and their mechanisms.

This introspective work is at the heart of mathematics. It implies deconstructing the mental images that you use without thinking and identifying where you can improve them. Done properly, this practice will reinforce your intuition day after day.

Mathematicians manipulate abstractions whose abstract nature they've forgotten, and that they prefer to call *objects.* They also like to say that these objects *exist.* By saying that, they don't necessarily want to take part in the old metaphysical debate that since Plato has argued over the reality of abstractions. They simply want to say that that's how they do math: by creating familiar ties to these objects, letting themselves imagine and manipulate them in their heads exactly as you would a banana.

To get to really know a mathematical object, you have to observe it for a long time, with intensity and detachment, with curiosity and open-mindedness. You need to take the time to play with it and create an intimate relationship, a relationship that takes place *outside of language.*

When Einstein said he was "passionately curious," when Grothendieck said he was "alone and listening to things, intensely absorbed in a child's game," this is what they were talking about.

13

Looking Like a Fool

When I started as an undergrad, I thought that mathematical creativity was reserved for people smarter than me. I thought that mathematical intelligence was innate and that everyone received a predetermined quantity. I was lucky enough to get a bit above the average. The geniuses got scandalously more.

I hadn't yet understood that mathematical intelligence was something you constructed for yourself. It's the natural byproduct of a physical activity that everyone is free to practice: mathematical imagination.

Mathematics is the science of imagination. Between those who allow themselves to imagine, observe, and manipulate mathematical objects and those who don't, there's an enormous divide. Over the years, this divide becomes monstrous, obscene, almost as monstrous and obscene as the divide between children with a room full of toys and games, and those who don't even know toys exist.

Contrary to popular belief, logic isn't the enemy of imagination. It can even be a close ally. The real enemy of imagination, that which blocks understanding and makes us feel like fools, is fear.

Fear is our real limitation. It concerns all of us, at every level, from the worst to the best, beginners to famous academics. We all have our blind spots, those words whose simple utterance fills us with terror because we've associated them with our deepest insecurities, our certainty that we're not good enough. We're petrified by the sign "Reserved for geniuses" that bars the entrance, forgetting that we put the sign up ourselves the day we told ourselves that math was too hard for us.

The saddest thing about the fear of math is that even though you know it's just in your head, it doesn't change anything. It's like fear of heights: you know it's only in your head, but all the same, you're still afraid.

Failed Conversations

Throughout my progress in math, I've experienced three big breakthroughs—three periods of liberation when, following a change in my psychological attitude, I felt the fear in me receding.

I spoke about the first two of these episodes in chapters 9 and 10. First, by paying attention to the dissonance between my intuition and logic, I chased away my fear of not succeeding on the first try. I allowed myself to imagine freely, even when I didn't yet fully understand something. Then, betting on an extreme mental plasticity, I started to address my fear of not being smart enough. If I observed the world with candor and sincerity, if I took the time to soak it all in, it was possible to overcome my limitations and become creative.

The third and most unexpected of these breakthroughs happened later, when I was in my thirties. I learned to chase away my fear of *being perceived* as not smart enough.

Up until then, despite an honorable beginning to my career and some initial success, I remained convinced that I wasn't a real mathematician. I attributed my success to luck. I told myself that I was an imposter, and that I would end up being found out. When I was teaching at Yale, I was having actual nightmares.

Our deepest fears are often social. For mathematicians, we're often afraid we're not as smart as the others, and that they'll see it.

I've seen this same fear in the eyes of most young mathematicians

I met. It's a natural enough phenomenon. I spoke in chapter 4 of the optical illusion that makes us underestimate the difficulty of math we really understand for the simple reason that it seems obvious to us.

There's a second factor that specifically concerns young academics. Normally, an academic is someone who knows things. When you become a professional mathematician, your social identity becomes that of someone who is smart. Except it doesn't at all work that way, and no one has warned you about it.

This misunderstanding can give rise to an aggressive form of imposter syndrome. I know people who have been overwhelmed by it, to the point where it permanently damaged their creativity.

Mathematics is a practice rather than knowledge. Mathematicians understand better than anyone the objects they're working on, but their mathematical intuition can never become omnipotent. Objects they aren't familiar with still raise difficulties. You can be an exceptional athlete, Olympic champion with the javelin, in peak physical condition, but that won't stop you from being crushed at tennis by a decent junior player.

In mathematical research, there is no position of authority. That creates disturbing, emotionally troubled situations that go against what is socially expected of us.

Here's a real-life situation that I personally experienced. You're supposed to be a bright young researcher. You've just gotten a prestigious position and you're an invited speaker at an international conference. At dinner, you find yourself seated next to a young doctoral student who is explaining what she's working on. You don't understand a word she's saying. You ask her a question. You don't understand her answer. Since you're stubborn, you risk telling her straight out that you don't understand. "No worry," she says, "let me reexplain it with a simple example; you'll get it right away." And there she reexplains it using different words, and you still don't understand anything she's saying.

The problem isn't with her explanation. The problem is with you. To understand, you'd have to start all over from the beginning, from the basics. Her work is related to a theory that you're supposed to have learned in graduate school, but which never made any sense to you. And, of course, you never dared tell anyone.

You're at the edge of what's socially acceptable. Your credibility is at stake. If you acknowledge how lost you are, you'll look like a fool. The social norm is to let it go.

This situation is typical of all failed math conversations, those we learn nothing from, those that serve only to reinforce our certainty that we're the worst of the worst.

Whatever your level of math, you know what I'm talking about. The vast majority of math conversations end with this feeling of malaise. They fail for this simple reason: you don't dare say how lost you are. You're ashamed, you feel ridiculous, and this idea gnaws at you and makes you incapable of listening. You think only of your own worthlessness. It's what keeps you from imagining and learning. You come out of these conversations feeling humiliated.

When I was thirty-two, I learned a social engineering technique to change the dynamics of these conversations.

I learned the technique directly from Jean-Pierre Serre, whom we spoke of in chapter 7, the person Grothendieck wrote to about his "ridiculous piece."

The lesson lasted five seconds flat and consisted of a single sentence. It's the most effective lesson in mathematical psychology that I've had in my life. I thought it over for months before I completely understood its scope. Thanks to this method, I've never again left a math conversation feeling humiliated.

I suspect there's a cause and effect: between age thirty-two and thirty-five, I underwent a formidable acceleration of my mathematical understanding. For the first time, I felt completely legitimate and at ease. I made spectacular progress in my research and I'm quite proud of the theorems I proved during this period.

The Art of Giving Math Talks

I'll of course share the technique Serre taught me, but first I have to explain the context.

Before new mathematical results are edited and published, they're generally presented in oral form at seminars and conferences. I've always enjoyed giving math talks, although the format can be intimidating, especially when it's at the blackboard. Talks usually last an hour, you're all alone with a piece of chalk in your hand, facing an audience of specialists who look at you impassively and may at any time interrupt you with questions. It doesn't leave a lot of room to bluff. But that's what's exciting.

I remember very well my first research presentation, in 1997 at the Isaac Newton Institute for Mathematical Sciences in Cambridge. I was a PhD student and I was absolutely terrified. To fight against my own insecurities, I chose to position the talk at the most elementary level possible, which demanded an incredible amount of preparation. I looked for ways to tell the story with the simplest mental images and the most natural connections.

In a way, I was trying to minimize mental energy—my own as well as that of the audience. A good metaphor is rock climbing: to climb up a cliff, you need to find a path and a series of actions that require minimal effort. You can succeed only if it's easy. If it's hard, you'll end up hurting yourself very, very badly.

This talk was a revelation for me. I understood that it was only by explaining it to others that I was able to really understand my own results. This is a well-known phenomenon, and mathematicians have a saying, that the only thing a math lesson is good for is to allow the professor to understand.

The best way for me to understand my own math is to imagine that I have to explain it to complete beginners. By playing the fool with myself, I end up finding ways to present my results as being obvious.

This minimalist approach became my presentation style, as opposed to the esoteric style and bluster behind which a lot of young mathematicians like to hide. I was initially afraid that the naïveté of my presentations wasn't doing me any favors. The risk was that people wouldn't take me seriously. The opposite happened. The simpler my talks were, the more intelligent people thought I was.

One day, I had to give a lecture at the Chevalley Seminar, a group theory seminar in Paris. I didn't have substantial new results to announce, but it was an opportunity to make a presentation even simpler than usual.

When I got to the room, fifteen or so researchers were there, along with a few students seated in the rear. A couple of minutes before the talk was to start, Serre came in and sat in the second row.

I was honored to have him in the audience, but I let him know right off that the presentation might not be very interesting to him. It was intended for a general audience and I was going to be explaining very basic things.

What I didn't tell him, of course, was that his presence was intimidating. Still, I didn't want to raise the level of my talk only to keep him interested. I just kept an eye out to see if he'd taken off his glasses, which would mean he was getting bored and had stopped listening. No worries there—he kept his glasses on till the end.

I gave my presentation as I would have without him there, speaking to the entire audience, especially the students seated in the back, whom I was pleased to see listening and looking like they understood.

It was a normal presentation, fairly successful, not very deep but well prepared, clear, and intelligible. At the end of the seminar, Serre came up to me and said—and here I quote verbatim: "You'll have to explain that to me again, because I didn't understand anything."

Looking Like a Fool

That's a true story, and it plunged me into a state of profound perplexity.

Apparently, Serre wasn't using the verb *to understand* the way most people use it. The concepts and reasonings of my talk couldn't really have caused him any difficulty. I'm sure he wanted to say that he understood what I had explained, but he hadn't understood *why* what I had explained was true.

It's a bit like with the sum of whole numbers from 1 to 100, where there are two levels of understanding. The first level consists of following the reasoning step by step and *accepting* that it's correct. Accepting is not the same as understanding. The second level is real understanding. It requires *seeing* where the reasoning comes from and why it's natural.

In thinking again about Serre's comment, I realized that my presentation had too many "miracles," too many arbitrary choices, too many things that worked without my really knowing why. Serre was right; it was incomprehensible. His feedback helped me become aware of a number of very big holes in my understanding of the objects and situations I was working on at the time.

In the years that followed, research into explanations for these various miracles allowed me to fill in some of the holes and achieve some of the most important results of my career. (However, some of the miracles remain unexplained to this day.)

But the most troubling aspect was the abruptness, the frankness with which Serre had overplayed his own incomprehension.

It takes a lot of nerve to listen closely to a presentation, then go up to the speaker, smile, and tell him that you "didn't understand anything." I never would have dared.

Why did he do it? I first told myself it must be one of the things you have the right to do when you're Jean-Pierre Serre. Then I realized that could also work the other way: what if this technique had actually helped him *become* Jean-Pierre Serre?

I decided to try it myself, just to be sure.

A few months later, at a conference, I found myself at a table next to a PhD student. During dessert, he started talking to me about what he was working on. Needless to say, I didn't understand a word he was saying. After dinner, I took him aside and said: "Explain it to me again, but very simply, very slowly. I don't understand anything about your subject. Assume that I have brain damage and can't focus my attention for more than a few seconds."

That made him smile, and he had the kindness to explain it slowly and calmly, starting from the beginning, with the basics in his field, that I should have known but up till now had never succeeded in understanding.

His explanation didn't have anything to do with what he had said over dinner. He didn't use the same words, and didn't even talk of the same things. It was as if he had two completely different ways of talking about the subject of his research. It was like there was the *tourist menu,* the official explanation he served up when he wanted to appear serious, and the *secret menu,* the simple and intuitive way he understood the things himself.

Because he was a student and I was an established scholar, I had higher social status. At dinner, he had tried to impress me by serving me the *tourist menu.* By overplaying my own worthlessness, I granted him permission to place himself on equal footing and just blurt it out.

Another benefit of Serre's technique is that it readily takes away the drama of all the stupid questions that you'd want to ask. Instead of asking them bit by bit, having the feeling that you're going backwards and losing your dignity with each passing moment of the conversation, it's much more comfortable to jump right in from the outset, admitting that sure, you're going to ask a lot of stupid questions, and you'll even be asking the same stupid questions over and over.

If you start up a math conversation, it's to learn something, not to be humiliated.

Sometimes you spend half the time reviewing the basics that you'd misunderstood, and sometimes that's all you do. At any rate, that's better than to talk about things you can't make any sense of. If the person you're talking with doesn't place themself at your level, and refuses to start with the basics and lead you by the hand, there's no use getting distraught. You've probably stumbled across an actual fraud, someone who pretends to explain math that is beyond their own comprehension. The real imposters are the ones without the syndrome.

The beauty of this approach is that by playing the fool you'll end up impressing people with your own self-confidence.

Refusing Fear

Serre's technique is simple and powerful. And it would seem everyone could do it. In theory, nothing is stopping you from looking people in the eyes and telling them smilingly that you didn't understand a thing and they need to explain it again from the beginning. Clearly, it's not a question of IQ.

Try it and you'll see.

It seems easy, but it's not. It might be hard to bluff and pretend that you understand. It's even harder to stop bluffing altogether and ask all the stupid questions that come to mind, without filter, without shame. Serre's technique is the social version of what we called in chapter 7 *the child's pose.* This requires a great command of your body and emotions, because we have the instinct to hide our ignorance.

And it's spectacularly harder if you worry that you might be written off as an actual idiot. This is why math is such an amplifier for all stereotypes and social insecurities. If you're part of a minority that lacks recognition and role models, if you're secretly convinced that your genes make you incapable of understanding math, or simply if being bad at math has become part of your social identity, it will be so much harder for you.

In the end, fearing that math is too difficult for you is a self-fulfilling prophecy. I have no easy fix to offer, apart from a few practical tips and the generic advice to keep pushing.

What Serre taught me was that it's better to be straightforward and direct rather than to beat around the bush. You might as well make fun of revealing what makes you ashamed and what you'd like to hide. Humor is the best weapon I know against fear. By pushing your own intellectual limitations to absurd levels, you can create a temporary zone of childlike freedom where any and all questions are allowed.

It's also essential to seek the right mentors. I've already spoken in chapter 9 of the video interview given by Pierre Deligne after he received the Abel Prize. It was an opportunity for him to share his vision of mathematics and recount some of the decisive moments of his career, including his first interaction with Grothendieck, before the latter became his PhD advisor.

Deligne, who was a young student at the time, went to a seminar given by Grothendieck, and was intimidated by his large silhouette and shaved head. During the presentation, Grothendieck spoke nonstop about "cohomology objects," a mathematical concept central to his work. But Deligne didn't understand any of it. At the end of the lecture he approached Grothendieck and asked him to explain what he meant by "cohomology objects."

It's a little like sitting through a lecture by Einstein and going up to him afterwards to ask what he meant by "relativity." More than a half century later, Deligne still admired Grothendieck's reaction: "That was really typical of him. Other people would have thought that, if I didn't know what this was, then really it was not worth speaking with me. That was not his reaction at all. Very patiently he [explained it to me]."

This patience and benevolence had a great impact on Deligne, and allowed him to flourish:

> He was extremely kind, one could ask apparently completely stupid questions. Being with him, I wasn't shy at all asking questions which would be completely stupid, and I've kept this habit until now. I usually sit in front of the audience attending a lecture, and if I have something I don't understand I'll ask, even if I would be supposed to know what the answer is.

These aren't just words. If Deligne takes the time to insist, it's because he knows how difficult it is. He's seen so many mathematicians fail at precisely this point, because they aren't able to attain the right level of candor and transparency. The most difficult thing in math is to overcome our shame or instinct for flight, our reflex for dissimulation. It's all a question of composure and physical engagement.

In his social media profile on MathOverflow, Thurston wrote something similar:

> Mathematics is a process of staring hard enough with enough perseverance at the fog of muddle and confusion to eventually break through to improved clarity. I'm happy when I can admit, at least to myself, that my thinking is muddled, and I try to overcome the

embarrassment that I might reveal ignorance or confusion. Over the years, this has helped me develop clarity in some things, but I remain muddled in many others.

Serre, Deligne, Thurston, Grothendieck: if these outstanding mathematicians all insist on the same point, it's not a coincidence. The struggle against our inhibitions and stumbling blocks is the essence of mathematical work.

It's normal not to understand. It's normal to be afraid. It's normal to have to struggle to contain your fear. It is, in fact, precisely what's at stake.

14
A Martial Art

Early in the year 1649 René Descartes received an invitation from Queen Christine of Sweden via the ambassador of France at Stockholm. She wanted him to come and give her private lessons.

Before accepting, Descartes wanted to make sure that she was serious. He told the ambassador: if this is only a whim, if the queen doesn't have the necessary motivation to truly learn, he wouldn't make the trip.

The year before, he'd wasted his time by accepting an invitation to Paris: "What disgusted me the most was that none of them showed any desire to know anything else about me but my face." It is the price of celebrity. He had the feeling that people wanted him not for his ideas, but "like an elephant or a panther, because of its rarity."

Three centuries before Einstein, Descartes was one of the first intellectuals to achieve the status of a rock star.

He ended up accepting Queen Christine's invitation and went to Stockholm, where he died of pneumonia at the age of fifty-three on February 11, 1650. When his body was being returned to France, his skull was stolen. It circulated on the black market for two centuries. The various owners engraved their names on it, as if the skull were endowed with magical powers that they could appropriate. The skull finally found its resting place in the Musée de l'Homme in Paris, where it is displayed next to the skull of an Australopithecus.

"The most evenly distributed thing in the world"

By now, we know the story by heart. It's the same story we've been repeating from page 1. It's the story of our refusal to believe that math is first and foremost an *attitude*, and our insistence that people who are good at it must have some kind of brain abnormality. And when an actual genius dares say otherwise, we cut his head open to see what's inside.

In 1637, before becoming a celebrity, Descartes had published an autobiographical essay, *Discourse on Method*, in which he outlined his intellectual journey. He revealed his work methods and told how he had become the greatest mathematician of his time. From the opening lines, the message is radically clear, and Descartes thinks he has no special talent:

> For myself, I have never presumed my mind to be any way more accomplished than that of the common man. Indeed, I have often wished that my mind was as fast, my imagination as clear and precise, and my memory as well stocked and sharp as those of certain other people.

What Descartes admits to is having a different way of looking at things, thanks to a particular method that he was lucky enough to stumble upon:

> I have fashioned a method by which, it seems to me, I have a way of adding progressively to my knowledge and raising it by degrees to the highest point that the limitations of my mind and the short span of life allotted to me will permit it to reach.

This method, as he describes it to us, is of a childish simplicity. It requires only one mental resource, the "good sense" that we're all endowed with. In other words, we all have the potential to become Descartes. In order not to leave any doubt, he opens his book with a sentence in the form of a slogan: "Good sense is the most evenly distributed thing in the world."

We never had the chance to have our discussion with Einstein. *Harvests and Sowings* is an obscure, almost unreadable text. *Discourse on Method,* on the other hand, is one of the most widely read and commented upon texts in the history of thought. How is it possible that nearly four centuries after its publication almost no one is aware of the existence of a *method* for becoming good at math?

Our collective inability to read Descartes is quite frankly extraordinary. We pretend to read it, we pretend to understand it, we pretend to find it important, but in reality we categorically refuse to take it seriously. Deep down we're all convinced that he's making fun of us.

This misunderstanding runs deep and explains much of our failure to democratize mathematics. But it is much broader than that. In fact, its extent is abysmal.

Indeed, Descartes didn't stop at mathematics. After developing his particular method, after practicing it with an iron discipline, after validating it through great mathematical discoveries, he set out to use it to reconstruct the entirety of science and philosophy.

The school of thought he founded is called *rationalism.* Our science and technology are direct descendants. Rationalism encountered limitations and pitfalls that Descartes didn't foresee, which we'll return to later. That doesn't take away from its success. The rationalist approach lives on in each of us, and whether we like it or not, we all know that rationality serves a purpose, just as we all know that math is extraordinarily powerful. Even if we have a hard time explaining why.

When we refuse to take Descartes seriously, it's rationality itself that we are refusing to understand.

Discourse on Method isn't a book of theory. It's a personal testimony, in which Descartes describes a number of mental techniques that he experimented with on himself. He affirms that these techniques allowed him to develop his cognitive abilities, build self-confidence, and make great discoveries. As proof of what he's saying, he accompanies his text with three scientific texts, including *Geometry,* a mathematical work so revolutionary that it remade our language and imagination (for the first time in history, Descartes used the letter x to designate an unknown).

Discourse on Method is a self-help book whose message is simple: we have the ability to construct our own intelligence and self-confidence.

The Secret Rationality

When you're a mathematician, you often meet people who say to you, "You're the rational one" or "You're the one who likes logic" or (worst of all) "You're the one who's good at math."

It's always a bad sign, because behind that there is always an underlying tone of "You're one of those weirdos," "You know nothing about life," or "You're the one on whom I'm going to pour out all the frustration I've built up through my school years."

Rationality has as bad a reputation as mathematics. And like the latter, it exists in two versions. The visible side of rationality presents itself in the form of established knowledge, science and technology, and well-structured and logically sound arguments. Schools spend a lot of time teaching it, with mixed results.

The flip side of rationality, its secret and intimate dimension, remains largely undocumented, as if we'd made a deliberate choice to cover it up.

In chapter II, I presented reason (and thus rationality) as a synonym for System 2 (mechanical thinking that follows rules and logic). It's a convenient shortcut because that's what it means to a lot of people. But it also results in major issues. By opposing rationality to System 1 (intuitive and instantaneous thinking), we're setting it in opposition to human understanding. It's no surprise then that so many people view it as dry and unappealing.

This version of rationality is a tough sell, but it doesn't stop some from trying to sell it. Most often they adopt a stance of superiority and disdain. "Be rational" is their way of telling us "Eat your vegetables," "Do your homework," "Respect authority," "Curb your desires," "Agree with me." They order us to be rational but they're incapable of explaining precisely what that consists of. They praise Descartes without realizing how much they're getting him wrong. It never even occurs to them that, with such pitiful ideas, Descartes would have long been forgotten, and none of his contemporaries would have admired him "like an elephant or a panther."

If you open *Discourse on Method* looking for a glorification of System 2, you'll be sadly disappointed. Descartes's great innovation was to put intuition and subjectivity at the heart of his approach to knowledge. He was distrustful of established knowledge and what was written in books. He placed little credit in authorities. He preferred to reconstruct everything by himself, in his head. His method closely resembles that of Einstein, Thurston, and Grothendieck. It is, of course, System 3, the slow and careful dialogue between intuition and logic, with the aim of developing your intuition.

Descartes was candid about it: his method was simply that of mathematicians. He described it without ever speaking of rationality or rationalism. These words didn't even exist in his time. They were invented later to characterize his approach. As for whether or not Descartes would pass today for a "rational person," I'll let you judge for yourself.

"The great book of the world"

René Descartes was born in 1596 in a small village in the center of France, whose inhabitants have since renamed it Descartes. The world Descartes lived in was very different from our own. In order to understand his thought, we must first understand his world.

The best way to ruin *Discourse on Method* is to see it as a world classic that might give you advice on how to succeed in school, or the work of an academic that explains how to do research that will win the esteem of your colleagues.

This isn't Descartes's message, and his life doesn't correspond to any modern stereotype of intellectual life. He never held an academic position, and he didn't make a living writing. He had been a brilliant student, eager to learn as much as he could, but passed a harsh judgment on what he had been taught: "As soon as I had finished my course of study, at which time it is usual to be admitted to the ranks of the well-educated . . . I found myself bogged down in so many doubts and errors, that it seemed to me that having set out to become learned, I had derived no benefit from my studies."

Descartes decided to turn away from scholarly "speculations" and learn directly from "the great book of the world," going wherever that might take him: "I spent the rest of my youth travelling, visiting courts and armies, mixing with people of different character and rank, accumulating different experiences, putting myself to the test in situations in which I found myself by chance."

A Martial Art

Descartes was pretty explicit about his great passion in life: "seeking truth."

It's easy to make fun of that, to sneer, to look condescendingly from the top of our postmodern heights, to pretend as if the notion of truth were a thing of the past. However, this is precisely where Descartes has the most to teach us.

As long as we conflate rationality and logic, as long as we reduce truth to its social and linguistic dimensions, as long as we see it only as a matter of consensus or authority, we completely miss the point of the Cartesian approach.

For Descartes, truth was a matter of life and death. He perfectly embodies this singular and powerful aspect of mathematical psychology: his relationship to the truth is physical, almost carnal:

> I constantly felt a burning desire to learn to distinguish the true from the false, to see my actions for what they were, and to proceed with confidence through life.

Descartes cared less for easy truths, those that are supposed to be true because tradition or such-and-such a person says so, or simply because they seem true. What interested him were solid truths, those that weren't going to change overnight, the ones you can rely on to become stronger and more confident, to make the right choices in life.

He approached truth as a martial art, an instinct you develop and that becomes embodied in action. Everything else—the philosophical arguments, the "opinions" of intellectuals with no skin in the game—was all just talk and of no interest to him: "For it seemed to me that I could discover much more truth from the reasoning that we all make about things that affect us and that will soon cause us harm if we misjudge them, than from the speculations in which a scholar engages in the privacy of his study, that have no consequence for him."

In this light, Descartes's uncommon obsession with fencing isn't that surprising. At age twenty, he overcame his self-confessed "dislike

[of] the business of writing books" and wrote a two-part treatise on the matter. The manuscript has since been lost, but a surviving summary shows us his precocious interest in the problem of mastery of the body.

If *The Art of Fencing* were published today, it might become an instant best seller, at least if the second part of the treatise lived up to its promise: "how you can always beat your opponent" if it's a competition of "two people of the same size, same strength, and same weapons."

But here again, our modern perspective might make us completely miss the point. For Descartes, fencing wasn't a hobby done on weekends at a club between well-meaning people. Nor was it a metaphor for intellectual jousting. Fencing was quite literally a martial art, an art of war.

At age twenty-two, confident in his body and his method, he started off in a career generally thought to be of little use in the development of the mind: he signed up as a mercenary.

The Rational Dream

During Descartes's time, the great minds of Europe were tormented by a deep question: does the Sun turn around the Earth, or is it the other way around?

It takes a real effort of the imagination to picture what "torment" might actually mean here. In today's world, there's no scientific debate that engages people's minds with the same intensity. There might be some people who debate whether the Earth is flat, but it would be greatly exaggerated to say that science is "tormented" by the question.

By challenging the traditional view that placed the Earth at the center of the universe, Copernicus had sparked much more than a scientific quarrel. He had forced the Christian world to consider this existential question: is the truth necessarily what is written in books, or do we, as human beings, have the ability to discover it ourselves?

On the night of November 10–11, 1619, at the age of twenty-three, while he was stationed at Neuburg an der Donau in Germany, Descartes had a series of three dreams.

The first was rather complicated and notably featured a melon someone wanted to give him and which, according to Descartes, represented "the charms of solitude."

In the second dream, he had the feeling of being struck by lightning. He woke up with a start and saw sparks all around him, as if the room were on fire. Here again he gave an interpretation: it was "the Spirit of Truth" that had come to take possession of him.

The third dream was what is called a *lucid dream:* in the middle of it, Descartes became aware that he was dreaming, and began to interpret his own dream while it was happening.

A dictionary appeared on the table. He was glad and told himself it might come in handy. But a second book drew his attention, a large collection of poetry that he was leafing through when a stranger came and showed him a poem. Descartes recognized the beginning of "The Pythagorean Yes and No" by the Latin poet Ausonius, and he began to look for it in the collection.

A bit later, Descartes realized that the dictionary was damaged. Then the man and the books disappeared. Without waking up, Descartes interpreted the dictionary as a symbol of science and the poetry collection as a symbol of philosophy and wisdom. The substance of the dream was precisely that: it is necessary to reconstruct science while being inspired by the techniques of poets who, through "the divinity of enthusiasm" and "the force of imagination," are able to uncover "the seeds of wisdom (which are found in the spirit of all men, like the sparks of fire in stone)."

Descartes thus invented rationality. Upon awakening, he was convinced that the Spirit of Truth had descended upon him to "open the treasures of all the sciences" by revealing that truth is not to be found in books, but in our heads. We have the ability to discover it ourselves, by the power of our thought.

For Descartes, evidence is the core criterion of truth. Not superficial evidence, the initial intuition that is often false, but evidence constructed through a deliberate and systematic effort of clarification, verbalization, and explanation, which aims to make things perfectly intelligible until they become fully transparent: "Things that we conceive of very clearly and distinctly are always true."

The True Mathematics

This revelation would guide the rest of his life. The following year, in 1620, Descartes abandoned his military career to dedicate himself to science. He began with arithmetic and geometry, "the easiest and clearest of all the sciences," where he made an auspicious debut with first-class results.

He saw in mathematics the basis of all knowledge, not only in the technical sense we've grown accustomed to (since the seventeenth century, mathematical formalism has become one of the basic tools of science), but also, and above all, in a basic, primordial sense, seated in the depths of human psychology.

For Descartes, the experience of understanding mathematics is the sole means of understanding what "understanding" really means. It has an intensity and bewildering uniqueness that acts on us like a revelation. Mathematics is a spiritual awakening. It teaches us to recognize the correct physical sensation, that which will guide us on the path to knowledge. And until we have personally encountered this crystalline and translucid form of truth, it is impossible to know what it means for a thing to be "clear and distinct," it is impossible to understand what Descartes is trying to tell us, and, according to him, it is impossible to enter into a real approach to knowledge.

Toward 1628 he began writing *Rules for the Direction of the Mind,* his first attempt to lay out his method. This text, which Descartes never chose to publish, prefigures *Discourse on Method,* which he would write nearly ten years later.

In it, he presents official mathematics as an "outer garment" that must be removed before its real substance can be accessed: "I would not value these Rules so highly if they were good only for solving those pointless problems with which arithmeticians and geometers are inclined to while away their time, for in that case all I could credit myself with achieving would be to dabble in trifles with greater subtlety than they."

Descartes contrasts what he calls the "true mathematics" with the "childish and pointless" stuff you find in textbooks. In his own effort to learn arithmetic and geometry, Descartes had grown frustrated at the disconnect: "In neither subject did I come across writers who fully satisfied me. I read much about numbers which I found to be true once I had gone over the calculations for myself. The writers displayed many geometrical truths before my very eyes, as it were, and derived them by means of logical arguments. But they did not seem to make it sufficiently clear to my mind why these things should be so and how they were discovered."

The ancient Greek philosophers granted a special place to mathematics. They made it the prerequisite of all philosophy and science. Legend has it that the following phrase was engraved at the entry to Plato's Academy: "Let no one ignorant of geometry enter."

For Descartes, this wouldn't have made any sense if the ancient Greeks had only known about the "childish and pointless" stuff. They must necessarily have been acquainted with the "true mathematics": "I came to suspect that they were familiar with a kind of mathematics quite different from the one which prevails today."

Descartes even has an explanation as to why this special kind of math wasn't passed on to us. According to him, the ancient Greeks intentionally chose to keep it secret, because it was too easy and too simple, and disclosing it would have damaged their intellectual prestige:

> I have come to think that these writers themselves, with a kind of pernicious cunning, later suppressed this mathematics as, notoriously, many inventors are known to have done where their own discoveries were concerned. They may have feared that their method, just because it was so easy and simple, would be depreciated if it were divulged; so to gain our admiration, they may have shown us, as the fruits of their method, some barren truths proved by clever arguments, instead of teaching us the method itself, which might have dispelled our admiration.

Reconnecting with the Truth within Us

Rules for the Direction of the Mind is a visionary text that anticipates the themes that we've addressed from the outset of this book. Descartes even formulated this profound truth, which seems so modern, that the principal objects stopping us are psychological stumbling blocks.

His approach is deeply meditative. To recover the path of this ancient knowledge, he invites us to reconnect with our primitive lucidity, the "primary seeds of truth naturally implanted in human minds [that] thrived vigorously in that unsophisticated and innocent age—seeds which have been stifled in us through our constantly reading and hearing all sorts of errors."

He notes that the difficulty is of an emotional and nonintellectual order. It arises from our social need to make believe, for others as for ourselves, that we understand what in reality we do not. It's a bit like the Fosbury flop: to adopt the correct position, you have to overcome the flight reflex that makes us believe that we're putting ourselves in danger. We have to rid ourselves of our false understanding and our tendencies to bluff.

This instinct for dissimulation particularly affects intellectuals: "Because they have thought it unbecoming for a man of learning to admit to being ignorant on any matter, they have got so used to elaborating their contrived doctrines that they have gradually come to believe them and to pass them off as true."

Our insecurity is such that we've abandoned the idea that real understanding is even possible. Because we refuse to believe that it could really be simple, we look for knowledge in the opposite direction, in the complicated and difficult.

Descartes realized how much his recommendations went against our instincts, so he took care to hammer home the message:

> We can have no knowledge without mental intuition or deduction.

> The whole method consists entirely in the ordering and arranging of the objects on which we must concentrate our mind's eye if we are to discover some truth.

> We shall be following this method exactly if we first reduce complicated and obscure propositions step by step to simpler ones, and then, starting with the intuition of the simplest ones of all, try to ascend through the same steps to a knowledge of all the rest.

> But many people either do not reflect upon what the Rule prescribes, or ignore it altogether, or presume that they have no need of it . . . as if they were trying to get from the bottom to the top of a building at one bound, spurning or failing to notice the stairs designed for that purpose.

> If in the series of things to be examined we come across something which our intellect is unable to intuit sufficiently well, we must stop at that point.

"My candour will be appreciated by everyone"

What the *Rules* didn't make explicit, however, was that Descartes was confronted with a major problem. This problem is central to his philosophy, and yet he was never able to resolve it: he had tried this method himself, he knew that it worked, but he could never explain why.

It's a problem that all mathematicians face, and one we've already discussed. It's difficult to talk about the unseen actions that we make in our heads. It's difficult to describe the method in tangible terms that others will make sense of. It's difficult to find a rational explanation for the fact that it works. The vocabulary that we possess to share our mental experiences is so poor that we rather quickly give the impression of saying whatever, nothing but smoke and mirrors.

And honestly, it's hard to take Descartes seriously when he speaks of the "Spirit of Truth" descending to take possession of him, or the "primary seeds of truth naturally implanted in human minds." It is, however, the only explanation that he was in a position to give. This lead him to adopt a *dualist* stance: he imagined a separation between mind and body. Our mind is of an immaterial nature, created by God in his image, and thus capable of attaining Truth as if by magic.

You can think what you like of this belief, but you can't take it for granted. At any rate, it's not at the level of rigor that Descartes aspired to impose upon himself: to accept as true only that which is so evident that it's impossible to doubt.

From a distance, you can see the extent of the problem. An explanation based on neuroplasticity was impossible to formulate in the seventeenth century. In Descartes's time, knowledge of the human body was rather limited. He himself thought of the heart as a furnace that served to heat the blood. At the time, medical practice still dressed itself in the guise of a carnival show before proceeding with enemas and bloodletting.

Among the great mathematicians in history, Descartes is far from being the only one to give a supernatural explanation for creativity. In the *Key to Dreams,* a mystic text that he wrote after *Harvests and Sowings,* Grothendieck says that God came to dream inside of him to show him the truth.

Srinivasa Ramanujan, whom we'll talk more of at the end of the book, said that his theorems were revealed to him by his family goddess Namagiri Thayar, a recurring character in his dreams.

I admit that these explanations have always left me a bit skeptical. We'll come back to them in the final chapters.

It's perhaps because he suspected he'd have a hard time convincing others that Descartes abandoned *Rules for the Direction of the Mind.*

Rather than directly stating his method, he chose to put it in practice. He pursued research in mathematics, physics, and biology. He traveled across Europe. He spent a number of years in Amsterdam, in the butchers' district, which gave him access to animal cadavers for dissection. Anatomy was one of his primary interests.

By the early 1630s, Descartes was readying an ambitious *Treatise on the World* that was to explain all natural phenomena.

But something happened that called the whole plan into question. In February 1633, Galileo was convicted of heresy and placed under house arrest for having defended Copernicus's theories.

Descartes, who never joked about his personal security, halted his research and the publication of his treatise. This same preoccupation with security had already led to his self-exile in the Netherlands, a Protestant nation and thus outside the reach of the Inquisition.

He resolved to pursue a partial publication, rather than a complete one, which would almost have guaranteed a conviction. He extracted from his treatise those chapters least likely to cause problems.

A Discourse on the Method of Correctly Conducting One's Reason and of Seeking Truth in the Sciences is Descartes's introduction to the collection he published, anonymously, in 1637. The three short treatises that followed this introduction (*Geometry,* the mathematical treatise we already mentioned, *Optics,* whose title is self-explanatory, and *Meteors,* which sought to explain phenomena such as wind, lightning, and rainbows) are presented as proof of the effectiveness of his method.

As opposed to *Rules for the Direction of the Mind,* written in an extremely preemptory and authoritative style, *Discourse on Method* begins in a strikingly modest fashion.

Descartes seemed to have realized that his story was a bit much and that people might have a hard time swallowing it. He didn't seem to completely believe it himself. From the opening pages of his book, he confessed that there was something of a contradiction. On the one hand, he was well aware of the utmost importance of his scientific work. In that respect, he didn't suffer from false modesty. On the other hand, it seemed to him that he wasn't particularly gifted.

He drew the conclusion that his success could only be explained by the method he had come across by chance. He recognized, however, that this might seem too good to be true:

> It is, however, possible that I am wrong, and that I am mistaking bits of copper and glass for gold and diamonds. I know how likely we are to be wrong on our own account. . . . So my aim here is not to teach the method that everyone must follow for the right conduct of his reason, but only to show in what way I have tried to conduct mine.

In short, he wasn't there to give us lessons. Like *Harvests and Sowings, Discourse on Method* is an autobiography, a testimonial that we shouldn't necessarily take at face value: "I hope that it may prove useful to some people without being harmful to any, and that my candour will be appreciated by everyone."

Visceral Doubt

Descartes's doubt is a bit like Ben Underwood's clicking: people find it so hard to believe that it works that they don't even bother trying, or they give up before it's started to work.

Outside of the mathematical community, I don't believe I've ever come across anyone who has really taken doubt seriously.

What a waste! A great mathematician takes the trouble to tell us how he got there despite being what he deemed as of average intelligence. He set down in black and white that his account has no theoretical basis but should be taken as a source of "examples worthy of imitation." Generations of students have been beaten over the head and forced to regard the *Discourse* as a philosophical treatise. And almost no one has taken the trouble to try it *for real.*

Before ending this chapter, let's take the time to clarify what Cartesian doubt really is, and the personal benefits everyone can get from it. After all, this is the best possible introduction to another key concept that up to now we've only begun to address: the concept of the *mathematical proof.*

At school we're taught that Cartesian doubt is a *methodical doubt.* It's a way of saying that the method is based on doubt. But this expression can lead to confusion. It's easy to take it the wrong way, thinking that you have to doubt in a methodical fashion. That's completely impossible, for the same reason that it's impossible to fall in love in a methodical fashion. You can only doubt with your gut. All Cartesian doubt is visceral.

Wanting to doubt in a methodical fashion is to confuse System 2, mechanical thinking, with System 3, the dialogue between intuition and logic.

Descartes wasn't against System 2. For example, he recommended making lists so that you don't forget anything. But the process of doubt doesn't belong to System 2. You can't doubt with words, you can doubt only silently, in your head. Doubt is personal and intimate. If you only pretend to doubt, if you don't go all the way, if you don't take the plunge, it's worth nothing.

In inventing doubt, Descartes positioned himself in opposition to official knowledge. He lived in a world where truth was still conflated with authority: truth was tradition, what was written in books. The sciences were still the heirs of the Aristotelean approach, over two thousand years old. That approach consisted of compiling and trying to structure a hodgepodge of things supposed to be 99 percent true (or that someone supposedly serious had said were 99 percent true, or 80 percent, or 51 percent—you never really knew).

For example, when Aristotle explained why the Earth was round, he amassed a bunch of mixed arguments taken from other sources. The more arguments there were, the more convincing it was supposed to be, up to the point where he quietly explains that there are elephants in Africa and elephants in Asia, therefore the two ends meet, and therefore the Earth is round.

To doubt is to give an argument the sniff test and sense that there's something off. It's allowing yourself to ask, "What? Really?"

Descartes's position is really quite simple: he states that what is probable, even 99.99 percent sure but not 100 percent, might be interesting, but from a scientific perspective it's worth nothing, because you can't build on it. (As we'll see, this is an extreme view that, in its purest form, is highly problematic. Modern science is in fact built on things that are very likely yet not 100 percent certain, and there are good reasons to proceed this way.)

Teaching Cartesian doubt is difficult because it's neither a knowledge nor a mode of argumentation, and is therefore impossible to evaluate. No one can doubt on a piece of paper. Doubt is a secret motor activity, an unseen action. To doubt something is to be able to imagine a scenario, even seemingly improbable, where the thing could be untrue.

Descartes asks us to doubt not only what others say, but also, and above all, our own certitudes. This is at the heart of his method, and it's where we have the hardest time following him. To place in doubt our own certitudes is like doing a high jump headfirst and on your back: our instinct tells us it's dangerous. We're afraid of exposing our physical vulnerabilities, of foundering, of falling into a bottomless pit where we'll no longer have confidence in anything. And we absolutely don't see what we stand to gain from it.

If you've never really experienced math, you'll have a tendency to believe it's a bottomless pit. The level of certainty demanded by Descartes seems impossible to attain.

Mathematics gives us examples of truths in which we can have absolute confidence. It's not just about superficial truths, like $2 + 2 = 4$, but also profound truths, truths that are extremely interesting and not at all immediately apparent. We'll give several striking examples in the next chapter.

It's only through a relentless confrontation with doubt that forces you to clarify and specify each detail until it all becomes transparent that you're finally able to *create* obviousness. Doubt is a technique of mental clarification. It serves to construct rather than destroy.

"An attitude of curiosity that excludes all fear"

Outside of mathematics, the level of certainty required by Descartes turns out to be impossible to attain for profound reasons that relate to how language and our brains function, and that have only recently been understood (we'll come back to this later).

That takes nothing away from the strength and potency of Cartesian doubt. The message of personal development in *Discourse on Method* goes well beyond the field of mathematics and the search for eternal truths.

This message makes sense only in light of Descartes's personality and his motivations. He hated "those sceptics who doubt for doubting's sake, and pretend to be always unable to reach a decision." His "burning desire," as we've seen, was the exact opposite: "to proceed with confidence through life."

His approach to doubt is closely tied to his taste for intuition, which he defined as "the conception of a clear and attentive mind, which is so easy and distinct that there can be no room for doubt."

Doubt thus corresponds to the shadow zones of intuition. To really doubt something, you can't just claim that you doubt, you have to sincerely believe that this thing might not be true. In order to do that, you have to construct an image in your head that shows there's a place for doubt. If you can't do that, you can't doubt—you're certain, like you can be with 2 + 2 = 4. But once you're able to imagine a scenario where the thing can be untrue, doubt immediately starts the process of reconfiguring your mental representations.

Because it mobilizes the imagination, the Cartesian version of doubt resembles techniques described in the previous chapters, except that instead of being concerned with numbers or geometric forms, it's concerned with truth itself.

Cartesian doubt is a universal technique for reprogramming your intuition.

It's therefore not surprising to find in Descartes's writing very similar advice to that given by Grothendieck and Thurston. He asks us, for example, for total physical commitment in service of our cognitive development: "We must make use of all the aids which intellect, imagination, sense-perception, and memory afford in order, firstly, to intuit simple propositions distinctly."

Descartes never speaks explicitly of mental plasticity. This notion would be conceptualized only many centuries after his death. But his account of the benefits of his method doesn't leave much room for ambiguity: "I believed that, in practising it, my mind was gradually getting used to conceiving of its objects more clearly and distinctly."

Descartes discovered that when we make a sincere attempt at introspection, when we're attentive to our cognitive dissonance, when we force ourselves to grasp our most fleeting mental images and put words to them, when we have the courage to face the internal contradictions of our imagination, when we have enough calm and self-control to look beyond our prejudices and see things as they really are, it has the result of modifying our mental representations, of making them more powerful, solid, coherent, and effective.

What Descartes discovered is a property of the human body.

The preeminence of the vocabulary of vision in Descartes is in this instance striking. When he states that truth is that which is "clear and distinct," you could almost say that he's giving it a neurological definition. His method evokes that of Thurston and Ben Underwood: it's a method for learning to see.

Practiced right, doubt can induce a state of profound comprehension that amazed Descartes, as it amazes all those who experience it. It's an experience that leaves you transformed, and that in itself is well worth the effort.

Doubt is not only the secret behind Descartes's achievements, it's also the secret of his incredible chutzpah. Seen in this way, *Discourse on Method* is a master class in self-confidence. His version of rationality is concrete, personal, rooted in our deepest aspirations. The whole point is to make us stronger: "This **assurance** is one side of a mindset, whose other side is an **openness to doubt:** an attitude of curiosity that excludes all fear as regards one's own mistakes, that allows us to detect and constantly correct them."

This last quote, taken from Grothendieck's *Harvests and Sowings*, perfectly summarizes this fundamental lesson from Descartes, and captures a unique aspect of the mathematical ethos.

Arrogant people who love being contradicted, show-offs who smile when you prove them wrong, dogmatists ready to change their mind in a heartbeat: I've encountered this singular attitude only among very good mathematicians.

15

Awe and Magic

If the mental actions of mathematicians were visible, research institutes would have glass walls. Passersby would stop to look, as they would at people who are kitesurfing or rock climbing. In high school, mathematics would be more popular than skateboarding.

In losing the possibility of imitation, we lose much more than our main learning method. We also lose our main driver of desire.

When you were a child, no one needed to make you want to ride a bike. No one had to convince you that it would serve you well later on in life, or that it would look good on your CV.

These questions never crossed your mind. You'd seen other kids riding bikes. You liked it and you wanted to do the same.

The principles of physics that govern the movement of bicycles have been known since 1687 and Isaac Newton's publication of *Philosophiae Naturalis Principia Mathematica,* the groundbreaking treatise in which he introduced both universal gravitation and the inertia principle. The bicycle wouldn't be invented until two centuries later. If you'd given one to Newton for his birthday, he would have refused to ride it. Most likely, he would have found the whole idea dumb and dangerous. He might even have gone so far as *proving* that it's physically impossible to keep your balance on a bicycle. But if you had shown him how, he would have been intrigued.

A Recipe for Mathematical Desire

If I could show it, the math that goes on in my head would intrigue a lot of people. But it serves no purpose to say so, since I don't have any means of showing it.

When I try to get people interested in math, I approach it differently. Rather than talking about what interests me personally, I choose the most accessible subjects and focus on re-creating the right emotional journey. After all, this is what got me interested in the first place: an emotional trigger, a very childish one, as when someone "dared" me and I didn't want to admit that I was scared.

I remember clearly how I felt the first time that I saw people kitesurfing. What I was seeing didn't seem at all possible. But at the same time—well, it was apparently possible. I ended up watching it for a long time.

Something similar intrigued me when I got started in math. It seemed too hard, too abstract, too incomprehensible. It didn't seem that doing math was humanly possible. And at the same time, it was apparently possible.

The difficulty of math, the initial shock, is only the first part of the emotional journey. The second part is the incredible feeling of wonder that arises from deep understanding, once you discover that not only is it possible, it is even easy. It was easy from the beginning, except that you couldn't see it.

Awe and then magic: this is a potent recipe for mathematical desire and a great complement to the sanitized math from the curriculum. Math isn't for the faint of heart. When we hide how scary it can be, we make it less desirable. If it wasn't for the awe, there would be no magic.

Measuring Infinity

A good subject for popularizing math is one that allows you to experience awe and then magic without getting burdened down with specialized techniques and language. In that respect, the discoveries of Georg Cantor (1845–1918) are perfect.

The concept of infinity goes back to the dawn of time and, for millennia, it had symbolized the unthinkable. You could talk at length about infinity, but only with a dignified composure and grandiloquent tone. It was a cool way of saying nothing while sounding profound. You were not allowed to talk about infinity in a casual and precise manner, in the same way you would talk about the number 5 or a straight line that intersects a circle at two points.

Being casual and precise when talking about infinity was like going to the Moon: it was the perfect example of something humankind would never achieve. Up until the day that Cantor found out he could do it. Even more incredible is that more than a century after such a spectacular discovery, most people haven't even heard about it.

When I meet someone who doesn't know there are many sizes of infinity, it's almost the same as meeting someone who doesn't know you can count beyond 5. It gives me a chance to share the news.

Take a grid that stretches to infinity. I'll only draw a bit of it, but you'll see what I'm talking about: a piece of grid paper that goes to infinity on all sides.

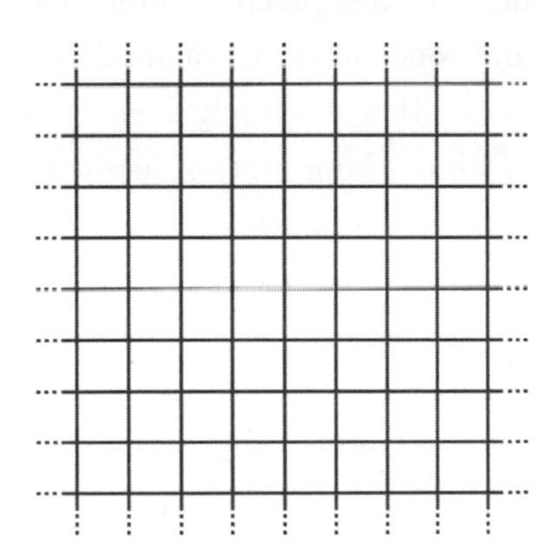

In this infinite grid, there are an infinity of white boxes. People generally understand this statement, which they find simple and concrete.

On an infinite straight line, there are also an infinite number of points. That's also generally clear to everyone. Everybody can picture this drawing as well.

Are there more boxes in the grid than points on the line? Is it the same number in each? Are there more points on the line than boxes in the grid?

I've heard people sneer at these questions. They were convinced that speaking about infinity was something reserved for mystics and theologians. Their typical reactions: "The question doesn't make any sense" and "Infinity doesn't exist."

You can't have it both ways: if infinity doesn't exist, then straight lines don't exist, or they only have a finite number of points. Mathematical abstractions exist neither more nor less than other abstractions we manipulate. Does the color red really exist? Do electrons really exist? Do justice and freedom really exist? In chapter 18 we'll talk about the insurmountable practical difficulties raised by a concept as pragmatic and concrete as that of an *elephant:* in a sense, elephants don't really exist. That absolutely doesn't stop us from talking about elephants, or from asking precise questions about them and providing answers to these questions.

Cantor realized that the language of sets allows you to give precise answers to questions about infinity.

The concept of sets is very old. It had been used since ancient times, informally, without anyone objecting to it or taking the time to look at it more closely. You could speak of "the set of houses on my street" or "the set of apples in front of you" or "the set of whole numbers" and everyone understood what it meant. The word was part of everyday speech and wasn't seen as a mathematical concept.

Building on the intuitive idea of what constitutes a set, Cantor created a simple and expressive mathematical vocabulary. His definitions are no more complicated than our theory of touch in chapter 8. Thanks to this vocabulary, it's possible to give a very precise meaning to the questions we asked above and provide a response as surprising as it is clear:

> Theorem: There are more points in the straight line than there are boxes in the grid.

Most strikingly, the whole thing, from the initial definitions to the theorem's proof, can be explained in under an hour to a curious primary school student. In other words, the solution of a problem that was deemed not just *unsolvable* but actually *unthinkable* was right before our eyes. It had been there since the beginning of time, less than an hour away from us.

I'm not joking: I've really explained it, over coffee, to the children of friends who'd invited me to lunch, and it really interested them.

Here's a rough outline of the proof, in a short and incomplete version. The infinity of boxes in the grid is said to be *countable:* you can number all the boxes with whole numbers (for example, by starting with a random box and numbering it as 1, then numbering the surrounding boxes as 2 through 9, and so on, numbering by successive nested squares up to infinity). Cantor discovered that, to the contrary, the straight line is *uncountable:* the infinity of its points is so large that you can't number them all using whole numbers. To prove that he used a process that today is called *Cantor's diagonal argument.*

Rather than trying to explain the details to you in writing (you can find them on Wikipedia), I'd rather let you find someone who can explain it to you in person. As we've seen in chapter 6, direct communication is spectacularly more efficient than reading. Follow the advice in chapter 13 and force yourself to ask all the stupid questions that come to mind: I guarantee that you'll have some.

Cantor himself was shocked by what he had discovered. He thought that it had been directly sent to him from God. As regards one of his most unexpected results (*there are as many points in a line as in a plane*), he admitted in a letter to one of his friends, "I see it, but I don't believe it!"

Cantor's results were so new and groundbreaking that he had to face the incredulity of his contemporaries. An influential mathematician labeled him a "scientific charlatan," a "renegade," a "corruptor of youth." When he submitted one of his articles to a leading scientific journal, the editor implored him to withdraw it because, he said, it came "about one hundred years too soon."

At the end of his life, undermined by all the controversies, Cantor sank into a deep depression. He spent his final years in sanatoriums and died in poverty.

His ideas finally prevailed. Since the early twentieth century, the concept of sets has become central in mathematics. For someone of my generation, trying to imagine doing math without sets is like trying to imagine life without electricity.

Grasping at Knots

When I want to explain what a mathematical proof is, what purpose it serves, and the special flavor of certitudes that you can construct *through the power of thought,* I like to use examples taken from knot theory.

In mathematics, a *knot* is a manner of joining the two ends of a string. For example, you can tie two ends of a string together in what is called a *trefoil knot.*

The string is presumed to be elastic and unbreakable: as long as you don't untie it, you can play with it as much as you like without changing the knot. For example, you can start with a string tied into a trefoil knot as above, and play with it to get this:

It's still a trefoil knot, just drawn differently. If you don't immediately see how you get from the first drawing to the second, if that gives you a bit of a headache, do not worry. It's normal. If it makes things easier, you can try it with an actual piece of string.

The simplest way to tie a string together is in a *trivial knot,* or an *unknot*, as follows:

In a way, the unknot is the "zero" of knots; it is a knot that isn't really knotted. You can draw it differently, for example:

It's clear enough visually that the string isn't really knotted, and that it's still the unknot. But there are other ways of drawing a trivial knot where it's not at all evident at first glance that you're dealing with an unknot. You can draw one, for example, like this:

Can you untangle this drawing in your mind, without the aid of a string or pen and paper? Despite my solid training in manipulating these kinds of things, it took me a while to figure out how to do it. If you can do it in a few minutes, without any special training, bravo! Once you get it the first time, it gets much easier to do it again.

To be honest, this example approaches the limits of my capacity for visualization. And these limits are greatly surpassed when it comes to mentally untangling a much more complicated drawing of an unknot, like this:

I don't know whether there are people really able to mentally untangle this kind of thing, to find it visually "obvious" that it's just the unknot. The idea terrifies me and even thinking about it gives me a headache.

It's precisely because it's hard to see that two different drawings represent the same knot that knot theory is an interesting subject.

Once you realize that there are an infinite number of ways, more or less complicated, to draw the same knot, you also realize that nothing guarantees that two knots drawn differently are in fact different. An initial legitimate question is, for example: *Is the trefoil knot really different from the unknot?*

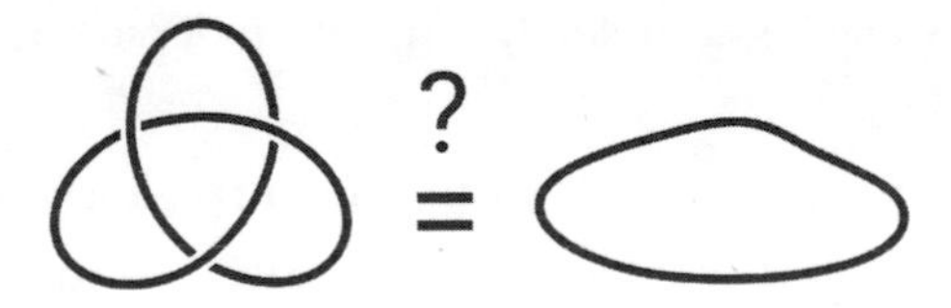

In other words: can you take a string knotted like a trefoil and twist it to undo the knot, without cutting the string, and lay it on a table so that it forms a circle?

If you try it, you'll quickly get the impression that it's not possible. Experimentally, you would say that a trefoil is not the same thing as an unknot.

I like this example because it's a perfect illustration of Cartesian doubt and the radical difference between an *impression* and a *mathematical proof.*

It's worthwhile to really try playing with a string for ten minutes, asking yourself the following question: what's your level of certainty that a trefoil knot really *is* different from a trivial knot—50 percent? 80 percent? 99 percent? 99.99 percent? Can you honestly say that there is no room for doubt?

I'll ask the question more bluntly: would you, *for real,* bet your life on it?

What guarantee do you have that there's not some winding road, some trick out of nowhere, that would allow you to go from a trefoil knot to the unknot?

It's like with those puzzles that seem unsolvable. If you have the solution, you know for sure there's a solution. But if you don't have the solution, you don't know whether there isn't one, or that you just haven't found it yet.

We all have the impression that a trefoil knot is different from an unknot, but the existence of complicated drawings of an unknot shows us that we can't always rely on our first impressions. Because a string looks like it's completely tangled up doesn't mean that it is really tangled up.

You could imagine that it's possible to untangle a trefoil knot and get back to an unknot by a series of maneuvers so complicated that no human has yet figured out how to do it.

At first glance, it thus seems impossible to be 100 percent sure about something. You'd have to consider the infinity of all possible ways of twisting the string. Even if you spent a billion years playing with the string, you could try only a finite number of combinations.

The beauty of mathematical reasoning is to be capable of manipulating objects as evanescent as knots and give answers of which you're 100 percent sure to questions that at first seem impossible to answer with that degree of certainty.

By "evanescent objects" I mean those objects that don't appear to be made to be rigorously manipulated by language. A knotted string is not like a whole number. It doesn't resemble something that you can tie up in equations. It's not something you feel able to capture in words.

With a trefoil knot, it does seem like the string is knotted and that you can't unknot it without cutting it. But you'd be hard pressed to say where exactly along the string the knot is situated. It's not at a specific place you can put your finger on. You sense the presence of the knot, but you can never really "grasp" it.

When I was a student, I was shocked to discover that it's possible to *grasp knots with language* and to give a complete proof, 100 percent certain, of this result:

Theorem: A trefoil knot is distinct from an unknot.

A proof of this theorem is outlined in the "Notes and Further Reading" section at the end of the book.

Stacking Oranges

It would be a lie to pretend that all mathematical proofs can be explained simply to nonspecialists.

The easiest problems to ask are sometimes the most difficult to solve. There are any number of problems that are easy to ask that we have no idea how to answer. There are even problems that are easy to ask that we know how to solve only with extremely complicated solutions, and for which there seems to be no easy answer.

Kepler's conjecture is a perfect example. It provides a tentative answer to the following question: *What is the best way to stack oranges?*

The great astronomer and mathematician Johannes Kepler (1571–1630) had an intuition for the solution in 1611 without, however, being able to prove that it was correct. As we've seen, a statement that you think is true but for which you can't give a rigorous proof is what mathematicians call a *conjecture.*

In Kepler's time, oranges were a luxury. He stated the problem in terms of cannonballs. But obviously that doesn't change the answer.

To be more precise, the question is about oranges presumed to be perfect spheres, all of the same size. If you tried to fill a space entirely with these oranges, how should you stack them to achieve maximum density?

With identical cubes, you can easily fill up the entire space without leaving any gaps, and get a density of 100 percent. But that's not possible with spheres.

At the market, fruit sellers usually stack oranges in a pyramid.

With this pattern, the density is about 74.05 percent: the space is filled about 74.05 percent with oranges, with about 25.95 percent of

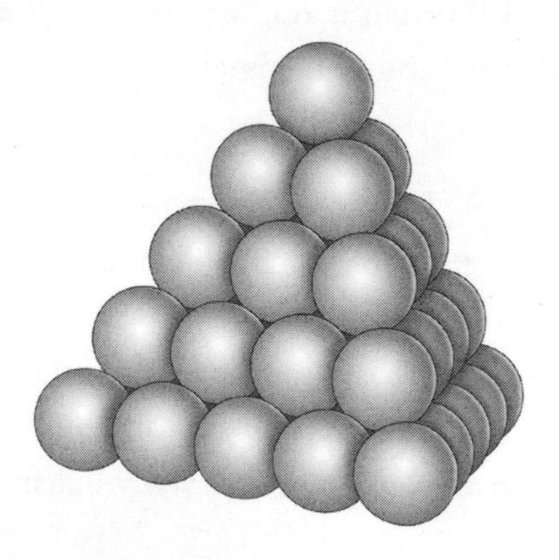

space lost between the oranges. Kepler's conjecture states that this is the maximum density and that there's no way of stacking the oranges that would be more dense.

Intuitively, that seems credible enough. But "credible enough" isn't what mathematicians call a proof.

For more than two centuries, no one made any progress. Gauss made the first breakthrough: in 1831 he proved that if you require that the oranges be arranged in a regular, repeating pattern, like atoms in a crystal, then the fruit seller's way of stacking the oranges is the densest possible.

It's a spectacular result but it doesn't totally prove Kepler's conjecture. A priori, it doesn't exclude that there might exist a very bizarre way, with no regularity, of stacking a hundred thousand billion oranges that would be denser than stacking them regularly.

Personally, I have no idea how you might approach such a problem. It makes my mind spin.

After Gauss's breakthrough, it would be another 150 years before Kepler's conjecture would be entirely solved. The first complete solution was given in 1998 by Tom Hales, an American mathematician born in 1958.

The conjecture thus remained open for 387 years. After such a long wait, habits are hard to break. Some people continue to say "Kepler's conjecture" for the result itself, whereas one really should say "Hales's theorem."

I'm completely unable to explain Hales's proof to a primary school student—or anyone else, for that matter—given that I've never looked at it closely and it would take me years of work to try to understand it.

In September 1998 Tom Hales submitted it to *Annals of Mathematics,* one of the most prestigious math journals. Before being accepted for publication, a scientific article is normally sent to one or two anonymous referees to check on its validity. Hales's proof was so difficult that *Annals of Mathematics* had to send it for evaluation to a committee of *twelve* referees.

It was even necessary to organize international conferences whose sole aim was to try to understand the proof. After four years, the chair of the refereeing committee affirmed that the committee members were "99% certain" of the validity of the proof: not bad, but still far from theorem grade. The article was finally accepted in August 2005, nearly seven years after it had been submitted.

An unusual feature of Hales's proof is that it is partly computer based: while an abstract mathematical reasoning allowed Hales to cover the general case, it left aside a finite number of possible exceptions that had to be studied separately. These millions of special configurations could be addressed only by brute force. It is this mixture of deep mathematics and massive computations that made the proof so difficult to assess. To this day no human has been able to prove Kepler's conjecture through the power of thought alone.

Kepler's conjecture bears on the piling of spheres in three dimensions, but the question of the optimal stacking of spheres could be posed in any dimension. As we mentioned in chapter 9, it is possible to do geometry in dimension n for any whole number n.

In two dimensions, the problem is easy enough to solve. A sphere in two dimensions is a circle. The problem thus becomes one of the densest way to arrange coins on a table. The optimal solution is shown in the figure.

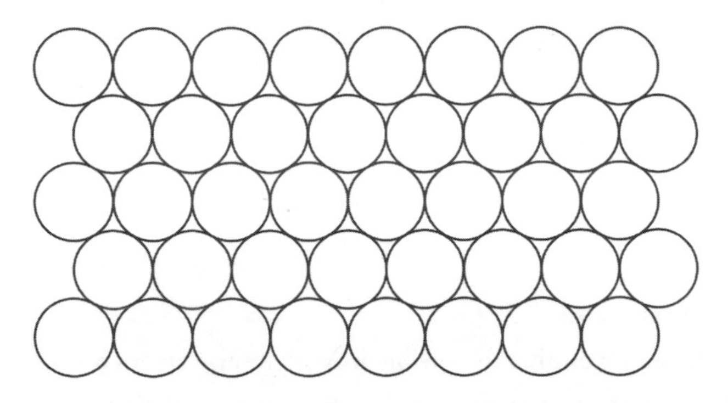

This result is in fact much easier to prove than the version in three dimensions.

Nothing is stopping you from trying to look *beyond* three dimensions. If you've never done geometry in higher dimensions, it's a bit intimidating to admit to yourself that there are people brave enough to take on these kinds of questions.

When it takes nearly four hundred years to solve a problem in three dimensions, you'd think you'd have to wait quite a while to see it solved for higher dimensions.

The solution still isn't known in four dimensions. Nor in five, six, or seven dimensions.

This is why the results of Maryna Viazovska, a Ukrainian mathematician born in 1984, created such surprise. In 2016, she began by solving the problem in eight dimensions using new and very elegant techniques. This already was a spectacular result. Three months later, along with four collaborators, she used a similar approach to solve the problem in twenty-four dimensions. For these breakthroughs, she received the Fields Medal in 2022.

These are for now the only dimensions above three for which the answer is known.

Why can we solve the problem in eight and twenty-four dimensions, but not in four or five? The explanation is that in eight and twenty-four dimensions, an unusual phenomenon takes place. There are exceptional mathematical objects, specific to these dimensions, that give rise to incredibly dense and harmonious ways of stacking the spheres. In twenty-four dimensions, the stacking is so dense that each sphere is in contact with 196,560 neighboring spheres.

I began this chapter by speaking of awe and magic. To characterize my personal reaction to the news that it is humanly possible to describe the best way of stacking spheres in twenty-four dimensions, *awe* is an accurate word. I haven't yet experienced the *magic* of understanding the proof myself, but I'm content enough with the joy that someone else did.

I imagine that for you the idea of twenty-four dimensions is already vertiginous in itself. The beauty of mathematics is that this vertigo can be overcome.

Nothing is stopping you from understanding what twenty-four dimensions really means, nor from understanding the usefulness of doing geometry in these types of spaces. There are practical uses for all this: the geometry of stacking spheres in twenty-four dimensions was most notably used in the data-transmission protocols for the Voyager 1 and 2 probes, sent by NASA outside of the solar system. The basics of geometry in higher dimensions is intellectually within your grasp. You can learn them in a few weeks. The most difficult thing is overcoming your fear.

Proving theorems like those of Tom Hales and Maryna Viazovska is obviously not within reach of everyone. However, getting to experience some of the awe and some of the magic is much more accessible, at the expense of an effort that in the end isn't so much.

Twenty-four-dimensional oranges can also be used to make another important point. Until recently, it was quite common to hear that women were biologically incapable of doing geometry. They were supposed to lack the ability to visualize objects in space, which made them incapable of even reading a road map. Next time you hear that kind of stuff, feel free to bring up Maryna Viazovska.

16
Hyperlucidity

My earliest memory of a deliberate and sustained effort to imagine something goes back to when I was seven. One night in bed, after having turned off the light and closed my eyes, I realized that with a bit of effort I could imagine seeing my favorite cartoon.

I didn't speak about it to anyone.

I remember like it was yesterday the amazement I felt and how I described the phenomenon to myself, in the words I used back then: it felt like I could "watch TV in my head."

I was able to visualize images and scenes I'd never seen. I was even able to imagine new episodes. It made a big impression on me. I loved it, so of course I continued doing it.

The transition between waking and sleep, as well as sleepiness upon awakening, have since that time played a central role in my intellectual development. In every project I begin, once it gets serious, once I really get interested, once I'm confronted with a real challenge, it starts to occupy that liminal space.

Transcribing my dreams was my first real attempt at writing. I was about seventeen when I started to get interested in this. At first, I found it too hard to directly write down my dreams, so I tried to record them while saying them aloud. My project was to collect them, like you would keep a journal or a photo album.

But something unexpected happened that forced me to give up the project. Night after night, as an effect of my trying to memorize them and put them into words, my dreams grew in richness and precision.

I was dreaming better and better. I was dreaming so well that it started to become annoying.

At first, I could remember only small fragments, bits and pieces of a single dream. But after two or three weeks, I was recounting five or six different dreams every day, each with a complete story and enough details to fill up a lot of pages with writing or long minutes of recordings.

It became overwhelming. The memories of my dreams were taking up too much space in my head and in my days. I felt like this exercise in introspection would end up consuming me.

Yet I continued to note them down, even after I'd stopped recording them. I wasn't looking for meaning in them. I just wanted to master the art of writing them down.

To me, this is the essence of writing. Starting from images and sensations and seeking a way to render them in words, to make them clear and solid. Transcribing the situations, what's at stake, how people and objects are positioned in space, their actions and movements. Describing what you see and feel as simply and faithfully as possible. Capturing the moods, the music, the smells, the textures. If you can do that, you can do anything.

Writing down dreams, in my experience, is the closest you can get to mathematical writing.

I'm always struck by the number of people who say they never remember their dreams. Some say they never dream. That's of course impossible. We all dream, each night.

Remembering your dreams isn't something you're born with. It's an ability you develop through practice. There are techniques to begin and techniques to get better. The more faithfully you learn to transcribe what you see, the more you see.

For a long time I kept a notebook on my bedside table, with a pen in the middle as a bookmark, to note down all my dreams and all the ideas that came to me during the night. I even taught myself how to write in complete darkness.

When I stop writing down my dreams I quickly lose my ability to remember them. When I force myself to write them down again, even if it's just a word or two, the ability gradually returns. Sometimes you have to keep it up for several weeks. The most difficult part is trying to recapture the first bit of a dream after a long period of not being able to do it.

As an adult, I've developed my own way of making use of the special state of mind just before falling asleep. Rather than *focusing* on subjects that preoccupy me, I've learned to simply let myself *be filled with them.* The nuance is subtle but fundamental. Focusing is thinking intensely, in search of solutions. It never works and it keeps you from sleeping. Being filled with something means contemplating it without a goal, in a decentered and disinterested manner. It's almost like dreaming.

I might be wrong, but it seems to me that this technique of falling asleep increases my chances of waking up the next morning with interesting ideas.

From Another Point of View

Among my exercises of geometric imagination, my favorite one is done during the phase of falling asleep.

Lying in my bed, eyes closed, I try to remember all the rooms I've slept in. I imagine being in the room: the size and orientation of the bed, the location of the walls and ceiling, doors and windows. I re-create the physical sensation of lying down in one of these rooms from my past. I try to experience it with my entire being. I choose the first room that comes to mind, then another, and another, and another in no particular order other than that of inspiration. The most interesting thing is when I recall a room that I had long forgotten.

I like this exercise because it's easy and peaceful. It's a great exercise for beginners.

As I said in chapter 1, I began early on to develop my intuitive perception of space and geometry, without seeing the least relationship to the math I was being taught at school.

It's so long ago that I can no longer say exactly when it started. The period when I was teaching myself to walk around with my eyes closed, memorizing the position of the walls and objects, was the same time I was imagining cartoons at night in bed.

I recall quite clearly my state of mind while I was doing those things. Early on, in kindergarten, I had to wear glasses because I was nearsighted. In my child's imagination, I believed that being nearsighted was just the first step toward becoming completely blind. I walked around with my eyes closed to train myself for the day when I could no longer see.

That's how I developed my capacity for visualization, and that's how I started learning geometry. Later on, I had the idea to use this capacity for visualization to train myself to be able, on command, to look at the world from another point of view.

I'm using the phrase "from another point of view" in a literal sense.

I was twelve years old when one day our art teacher asked us to draw our pencil cases. I drew mine *seen from inside,* with gigantic pens deformed by the perspective. No one had taught me to draw using perspective, and yet it seemed natural to me: I was simply drawing what I saw in my head.

I was quite proud of this drawing and I remember it made a big impression on my classmates.

Around the time I was fifteen we'd studied solid geometry in class, that is, geometry in three dimensions. It was only then that I realized my capacity for visualization differed from that of my friends. It's perhaps the most surreal memory of my school years.

The class content and the exercises seemed to me so completely useless that they didn't even make sense. It seemed like we were back in kindergarten. If the teacher had simply put one finger in the air and asked us how many fingers we saw, and we only had to answer, "One," it would have made exactly the same impression on me.

I got the best grade without understanding how it was possible not to understand. But I knew from my friends that solid geometry was a frightening subject for some of them. It gave them such difficulties that they were embarrassed to talk about it.

I had no reason to suspect that my ability to *switch viewpoints* was so unusual. I was completely unaware of the extent of my geometric over-training.

If you want to practice *switching viewpoints,* here is a good exercise:

1. Choose a random reference point around you, for example, the corner opposite from you in a room, or the window of a house when you're walking in the street.
2. Try to imagine what you'd see if you were looking in your direction from this reference point.

It's not a binary exercise in which there are those who can and those who can't do it. The exercise is hard for everyone. At first, it seems like you'll never be able to see anything. But you necessarily see something: a vague shape, a shadow or a blotch of light, or just a confused and fleeting idea. You have to start there. The goal is to increase the clarity and the resolution of the picture, while trying to maintain its presence for as long as possible.

I've tried a number of variations of this exercise. I've even tried completely stupid things, things that were entirely undoable, just for fun.

At a time when I often took the train from Paris to La Rochelle, I tried to visually memorize all of the countryside, always from the same side of the train (the west side), and to reassemble it into a single image in my head. The resulting image would have had to be huge: 150 feet high and 300 miles long. It would have meant creating an image 10,000 times wider than it was tall and filling it entirely.

I never succeeded.

Seeing the Unseen

After having used my imagination to "see" the world from other points of view, I got into the habit of systematically trying to "see" all the things that were supposed to be there, even though they weren't visible.

Let me explain this with an example, that of soap bubbles.

A soap bubble is, in essence, the same thing as a balloon. The soapy water forms an elastic membrane, which is inflated by the internal air pressure. This is why the bubbles are spherical: to enclose a given volume of air, the sphere is the shape with minimal surface area.

When a bubble forms, you can see it undulate for a few seconds before it takes its spherical form. During this initial undulation, it's quite easy to "sense" that the surface of the bubble is elastic. The bubble undulates slowly, like a balloon filled with water. We're familiar with this movement, and as it takes place it feels that we can "see" the physical properties of the bubble.

I've trained myself to continue to "see" the elasticity of the bubble once it has taken its spherical shape. I've trained myself to "see" that the air inside the bubble has a higher pressure than the air outside the bubble.

I'm placing quotation marks around "see" because I know that I can't really see it with my eyes. It's visual, and yet at the same time it's not really visual. I wouldn't be able to draw it. It's just a weird sensation that is located within my visual field, like if something had been highlighted. In a sense, you could say that it's a hallucination. But it's an educated hallucination, one that's constructed and controlled.

When I look at a bridge, I see stress lines. I see which parts of the bridge are subject to compression, and which ones are subject to tension.

I can call up these perceptions on command, with a bit of concentration. They help me experience and understand the world.

You have similar sensations. You can "see" that a rope is stretched too tightly and is about to break, or that a balloon is overinflated and is about to burst. You've learned to "see" the tension of objects as if it was another layer of augmented reality, supplementary information embedded in your visual field, exactly like a color but located in another layer of reality.

These examples might seem too ordinary. Educating yourself to see more and more of these types of things might seem childish and vain. Nevertheless, I have the feeling that it's this approach, applied systematically throughout my life, that has solidified my scientific understanding and allowed me to produce original mathematical work.

That which I'm unable to "see" or "feel," even when I know it should be real, retains a special status for me. I don't ignore the external information I get through language, but I treat it as a *hypothesis,* without entirely believing it.

If I don't see why something should be true, I'm wary of it. This kind of information can stay in this intermediate status for a long time. Maybe for a few hours, days, weeks, years, or even decades.

Sometimes I suddenly "understand" things I was taught when I was a child, and that had since then remained in this intermediate status.

In my class of high school geography, we were taught that deforestation led to soil erosion. This information, which I received through language, without visualization and without really understanding it, went completely over my head. It didn't interest me, and it didn't convince me.

A decade later, I finally understood. I remember the moment quite well. I was at a math conference, during a presentation where I was getting bored because I couldn't understand anything. I looked out the window at some trees. I played at visualizing the scenery in its totality. I imagined the trees in their entirety, down to their roots. And all of a sudden, it all made sense.

Visually, it was striking. The network of roots formed a structure that retained the soil and the rocks, like rebar in reinforced concrete, creating a composite structure with incredible strength. This explained why the soil wasn't sliding down along the hillside. It also explained another phenomenon I'd seen before and that had struck me: of how unbelievably difficult it was to remove a tree stump.

Another example: I knew that planes were able to fly, but a part of me continued to refuse to believe it, because it didn't seem normal, and I couldn't intuitively understand how it was possible.

This part of me came out when I was seated in a plane accelerating down the runway for takeoff. A little voice inside me whispered something like, "It's a joke, this thing can't fly, it's way too heavy, we'll never take off."

I suspect that many people have heard this little voice inside their head, at least once in their life. But they don't dare admit to it, afraid that they'll look like fools.

We're socially conditioned to find it normal that airplanes can take off, and to laugh at the idea that someone could doubt it. But how many of us really know how planes manage to fly?

It's not foolish to have doubts about the ability of planes to take off. It's simply proof of good sense and independence of mind.

My doubts never stopped me from boarding a plane. I didn't go so far as to deny the evidence. I could see that planes were able to fly. It was something that I accepted, somewhat reluctantly, because I never had been offered a choice. I accepted it from a practical standpoint but not from a sensory standpoint.

Of course, there are many more things I have to accept in this way than there are things that I'm really capable of understanding.

It's only been a few years since I've completely accepted the idea that airplanes can fly. I learned how to feel it physically. In order to do so I had to learn how to feel the density of the air and the phenomenon of lift. I had to find out that airplanes are much lighter than they appear to be. I had to learn to feel the wings, the way that they lift up and bend, their internal structure, how they are attached to the fuselage and why they don't break.

My intuition ended up finding all that normal. Planes became like an extension of my body.

In My Spine

Having confronted difficult mathematical subjects and made a career of it, I developed a lot of confidence in my geometric intuition.

It comes into play notably about subjects most people don't recognize as being geometric in nature. For example, when I learned of the discovery of *Homo floresiensis,* then Denisovans, two extinct spe-

cies of humans that cohabitated with ours not so long ago
that surprised. It kind of felt natural, for reasons owing to m
ition of what I might call the "geometry of the tree of life" and
ometry of fossil discoveries."

This doesn't make me a magician. My intuition is human an
fallible. I've only pushed to its extremes an ability naturally present in all of us.

What mathematics taught me was that it was really possible to get ahead and move forward in life without giving up the childlike desire to remain down to earth, to accept only what is concrete and evident.

That came as a surprise. I invented on my own these exercises of visualization and imagination, without expecting that they would have so much impact on my life.

In mathematics, as in many other fields, creativity is simply the ultimate form of understanding, which itself is but a natural product of our mental activity. It emerges when we force ourselves to continue looking at things that intimidate us until they finally become familiar and obvious.

There are many ways of grasping mathematics. We all start off with our strengths and weaknesses. Geometry has been one of my strong points since childhood and I have a long history of visual imagination.

I also have my weaknesses. For example, I never really was interested in numbers, and never spent enough time thinking about them. That being said, through repetition and habit, I did manage to develop a degree of familiarity with them. I understand them up to a certain level. But I've never felt myself capable of being creative in number theory or in arithmetic. Like a tennis player whose backhand is relatively weak and who shifts to play his forehand, I've fallen nabit of avoiding numbers whenever possible, and looking metric interpretations of quantitative statements.

r my primary interest had been with numbers, I would certainly e developed a much stronger personal bond with them. This inuition might have been primarily visual, or of an entirely different nature. In the end it doesn't matter that much. All mathematicians approach mathematical objects intuitively, but intuitions come in many shapes and forms.

My weakest point is my inability to find my way in complicated notations, to follow without stumbling reasonings that contain a lot of symbols and formulas. Entire areas of mathematics put me off because of this, especially analysis. It's only gotten worse over time as I've lost patience with it.

In certain subjects that at first I wasn't very good at, I've managed to get better later on. To understand the abstract structures of algebra that caused me so many difficulties when I was twenty, I've developed a particular form of sensory intuition.

These are powerful sensations but impossible to put into words. I grasp certain mathematical concepts through nonvisual motor sensations, tensions and force fields within my own body, as if I could transport myself and experience these objects *from the inside.*

I feel this math in my neck and in my spine.

I know these aren't the right words, but I can't think of better ones. Here is an exercise of the imagination that helped me a lot when I was trying to improve in these subject areas. I looked at an object, for example, a bottle of shampoo sitting in the bathroom, and asked myself the following question: if my body were shaped like the bottle, how would it feel physically?

By practicing the same exercise with mathematical objects, I was able to understand them. When I was thirty-five years old, this technique led me to experience the most intense creative period of my career.

It all started with a casual observation. One day, I noticed that a question I was asking myself about the geometry of certain braids in dimension 8 could easily be translated into the language of category theory. The unexpected link between two very different intuitions offered a novel way of looking at problems I'd been struggling with for years.

It was as if I'd built a bridge between two regions of my brain that up until then hadn't communicated with one another. There was a big jolt of comprehension, then a series of smaller aftershocks. But it was just the beginning. My mathematical imagination was about to undergo a massive reconfiguration.

Each day I woke up with new ideas. Some of them were tied to the problem I wanted to solve (a conjecture dating from the 1970s) but others took me in completely different directions. I thought they were beautiful, but I had to quit following them up because it was impossible to explore so many leads at the same time. It was just too much. I tried to take notes but my understanding progressed faster than my ability to write it down.

This state of hyperlucidity lasted six weeks, the time that I finished proving the conjecture.

I wasn't able to sleep. I was exhausted. Once I woke up at 4 in the morning with the urge to look at a book I'd bought ten years earlier (the second volume of *Representations and Cohomology* by Dave Benson), which, at the time, I had barely opened. I found the book on a shelf, grabbed it, sat on the floor, and read a hundred pages in one go, as if it were a comic book. I'd never been able to read a math book like that before. If I was able to read it so fast, it was because I already knew what was written down: it was as if I had just seen it in my dream.

Throughout these six weeks, I had the feeling of understanding more new mathematics than everything I'd understood in the previous twelve years, since I had begun as a PhD student. It was going so fast that I was seasick. It was physically overwhelming and I could no longer cope. I was hurting. I wished that it would stop, so that I could get some rest. But it wouldn't stop. It was like mathematics had taken control of my brain and was thinking from inside my head, against my will.

For the first time in my life, I realized that extreme math was a dangerous sport.

17

Controlling the Universe

If you were to think of an archetypal young math prodigy, I have a pretty good idea of what would come to mind.

It undoubtedly won't be the class clown who spends his time partying with friends. Nor a well-adjusted character, good at relationships, resourceful, and easy to live with. Nor the one who just takes it easy.

Perhaps, instead, you'd imagine someone like Ted Kaczynski.

Ted Kaczynski was born in Chicago in 1942. His exceptional math skills were soon spotted by the school system. After scoring 167 on an IQ test, he skipped a grade. A few years later he skipped another. He was accepted into Harvard in 1958, when he was sixteen years old. He completed his PhD in mathematics in 1967 and became the youngest assistant professor at Berkeley.

Apart from his academic success, Ted led a sad and lonely life. Those who knew him in his youth described him as emotionally impaired, incapable of communicating or building genuine relationships.

At Harvard, students he shared university housing with recalled two details: his habit of playing the trombone in the middle of the night, and the smell of rotting food that came from under his door.

When Ted was fifteen, he had such difficulties socializing with the other teens that he preferred playing with the friends of his younger brother David, who was eight.

David Kaczynski remembers his admiration for the intelligence of his older brother, whom he loved, but also his surprise at his strange behavior. What kept Ted from making any friends?

One day when David was eight or nine he asked his mother, "Mom, what's wrong with Teddy?"

The Spectrum of Oddness

From the beginning of this book I've said that we need to get beyond the stereotypes of math and mathematicians. I haven't changed my mind. But that doesn't mean that these stereotypes don't exist, or that they come out of nowhere.

It's no secret. When you get to know the mathematical community, one thing that immediately strikes you is the number of "odd" characters. "Odd" is a nice way of putting it. There are degrees of oddness. Some mathematicians are a bit odd. Others are quite frankly odd. Still others are spectacularly odd. In some cases, it seems to go beyond oddness. Only one word seems suitable: insanity.

It is in fact a recurring conversation topic among mathematicians. Everyone has dozens of stories to tell. These stories can be extremely funny, but they're so caricatured that they're hard to believe.

Among all the stories whose authenticity I can vouch for, let me share just one: I once had dinner next to a mathematician who wandered about with a garbage bag filled with train schedules that he knew by heart. The best way to start a conversation with him was to ask the options for traveling from New Haven to Philadelphia on a Sunday afternoon.

Getting beyond the stereotypes means keeping in mind that the majority of mathematicians aren't like that. It's entirely possible to do mathematics at the highest levels and still be "normal." You can be well rounded socially, warm, and open to others. You can even be a charismatic leader. Even more than oddness, humor is a fairly typical trait among mathematicians.

That doesn't mean that oddness doesn't occur, often spectacularly, notably among some of the most prominent mathematicians in history. In chapter 7 we spoke about the life choices of Alexander Grothendieck, his extreme solitude and asceticism.

Another striking example is Grisha Perelman, whom we've already come across in chapter 10. Born in Leningrad in 1966, he is famous for having proven in 2003 the conjecture formulated by Poincaré in 1904. It was a massive achievement whose scope is hard to conceive.

Perelman not only refused the Fields Medal, he also turned down the million-dollar prize awarded to him in 2010 by the Clay Mathematical Institute (the Poincaré conjecture figured on its list of the Millennium Prize Problems, the seven problems deemed to be the hardest and most pivotal for the future of mathematics).

In 2005 Perelman resigned from his position at the Steklov Institute. He doesn't give interviews and it's difficult to know what is going on inside his head. His personality is so intriguing that there are stolen photos circulating on the internet, along with fake interviews and crazy rumors.

He most likely never really spoke the phrase often attributed to him: "What would I do with a million dollars when I can already control the universe?"

Another statement, on the other hand, comes from a reputable source: "Money and fame don't interest me. I don't want to be put on display like an animal in the zoo. I'm not some mathematical hero. I'm not even as brilliant as all that, which is why I don't want everyone looking at me."

We can't but admire Perelman's creative intelligence, his mental prowess, his unfailing determination, and his incorruptible nobility of mind.

And at the same time there's something deeply troubling about his story. It comes across as a feeling of malaise. Is that really what genius is supposed to be? Does it always have to end this way? Is it really impossible to reach out to others and find a means of communicating with them?

We can't help feeling that something isn't right. We think that if Perelman refuses to cut his fingernails and, at nearly sixty years old, he continues to live with his mother in a small apartment in Saint Petersburg, it's because something's wrong with him.

It's possible. But what do we know? Perelman is one of the most brilliant minds on the planet. He's never done anyone any harm and has the right to live as he pleases. We're in no position to judge him.

Is it mathematics itself that makes people "odd"? I don't think so. I'd say rather that math is welcoming to people who are already a bit "odd."

It's one of those rare careers where it's possible to accomplish great things even if you can't fit in. For people already a bit "odd," "different," not at ease in society, it can be a path toward socialization and fulfillment. (I tend to put myself in this category.)

To most people, math isn't really that dangerous.

There is, however, one major contraindication, a specific context where math can have catastrophic side effects. If you judge by the examples that occur throughout history with troubling frequency, mathematics seems, in certain cases, to feed and amplify a particular mental pathology: paranoia.

Heart of Darkness

On the spectrum of oddness, Ted Kaczynski ranked near the top.

Creativity doesn't come in proportion to oddness. And precocity doesn't always foreshadow a brilliant career. Ted Kaczynski was never a great mathematician. His meager scientific output doesn't have any particular merit.

On June 30, 1969, at twenty-seven years old, Ted Kaczynski abruptly resigned from his position at Berkeley without giving any explanation. Two years later, he went to live in a cabin he built himself in the forest in an isolated part of Montana.

He chose to live alone, without running water or electricity.

Beginning in 1971, in his journal, he talks about his criminal intent: "I emphasize that my motivation is personal revenge. I don't pretend any kind of philosophical or moralistic justification."

His discourse changed over time. Later on, Kaczynski would claim that he wanted to start a revolution. But throughout he maintained a feeling that seemed visceral for him: a hatred of the scientific establishment and bureaucracy, which he saw as attacking his personal freedom and freedom in general: "My ambition is to kill a scientist, big businessman, government official or the like. I would also like to kill a Communist."

Over the years, during long solitary walks through the forest that he came to identify with, Ted Kaczynski developed his plan. He would get his revenge on industrial society for all that it had made him suffer, and for the violence it inflicted on the trees that it heartlessly cut down to build new roads.

With method and determination he fell further and further into an unbelievable murderous web.

He wasn't captured until 1996, after the longest and most costly investigation in the history of the FBI. It had lasted more than seventeen years and employed up to 150 full-time agents.

In 1998, Ted Kaczynski began serving eight consecutive life terms without the possibility of parole at the super-maximum facility in Florence, Colorado, the most secure prison in the United States. His co-detainees included the likes of Zacarias Moussaoui, one of the 9/11 terrorists, and El Chapo, the Mexican drug baron.

He died in 2023 in an apparent suicide.

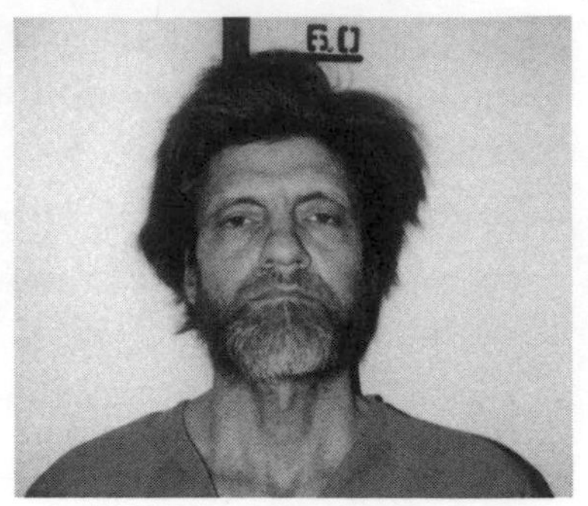

His story is bleak but it needs to be told. It can teach us something fundamental about rationality: its power, limitations, dangers, and the good and bad ways to use it.

Unabomber

On November 15, 1979, American Airlines flight 444 left Chicago for Washington, DC. Midway through the flight the passengers heard a muffled sound. The cabin filled with acrid smoke and the oxygen masks fell. The smoke was so dense that it got inside the masks.

The pilot was able to make an emergency landing. Twelve people were hospitalized with smoke inhalation. On the ground, the initial findings left no doubt: there had been a bomb in the plane which, if it had worked correctly, would have obliterated it midair.

A link was quickly established with two booby-trapped packages left on the campus of Northwestern University, near Chicago.

This was the first in a very long series of similar incidents.

On June 10, 1980, Percy Wood, president of United Airlines, was seriously injured by a package bomb sent to his home in Lake Forest, a suburb of Chicago. In 1981, a bomb was defused on the campus of the University of Utah.

The attacks continued until 1995. Sixteen bombs resulted in three people dead and twenty-three wounded. The targets were universities (a building in Berkeley was targeted twice) and businesses having to do with aviation, industry, or technology (among them the Boeing offices, electronics stores, and a lobbyist for the timber industry).

In this indecipherable case of serial terrorism, the investigators followed up on the slightest of clues.

Although some devices carried the initials FC, which was later revealed to stand for "Freedom Club," the FBI remained convinced there was only a single person behind the attacks. Agents constructed a psychological profile of a man fed by a fierce hatred against universities and aerial transport. He was also obsessed with wood and forests. This fascination manifested itself in the choice of targets (the timber industry, Percy *Wood,* Lake *Forest*) as well as in the materials used to make the bombs. Some contained pieces of bark; others were camouflaged to resemble logs.

The FBI and the media dubbed him the "Unabomber," short for "University and Airline Bomber."

As for the bombs themselves, they posed a lot of difficulties for investigators. Ordinarily, it's possible to make each fragment left at the scene "speak": by analyzing something as small as a nail, they might be able to identify the manufacturer and places where it might have been sold.

The problem was that the nails used by the Unabomber were all made by hand. There were no clues to follow, no fingerprints. The components of the bombs had all been carefully sandpapered.

The terrorist was someone patient and meticulous who was not afraid of building everything from scratch.

The investigators had to wait until 1995 for an opening in the case. That year the Unabomber sent a typed manuscript to the *New York Times,* the *Washington Post,* and *Penthouse* along with a letter in which he said he would give up his attacks if the text was published. Following the recommendation of the FBI, the *New York Times* and the *Washington Post* published it on September 19, 1995.

The Unabomber manifesto, "Industrial Society and Its Future," is meticulously constructed and argued. Constituting paragraphs numbered from 1 to 232, it's a radical critique of modern society and the hold that technology has on our lives. According to the Unabomber, there's nothing worth saving, so the entire system needs to be torn down.

The statements are at times relevant and at other times absurd. Certain passages betray preoccupations that you wouldn't expect from your typical terrorist: "It is true that one can ask serious questions about the foundations of scientific knowledge and about how, if at all, the concept of objective reality can be defined. But it is obvious that modern leftish philosophers are not simply cool-headed logicians systematically analyzing the foundations of knowledge."

When David Kaczynski read this passage, he was struck by the phrase "cool-headed logicians." He remembered a letter in which his brother Ted used exactly the same words to say exactly the same thing.

Suspecting that his brother was the most wanted criminal in the United States, David Kaczynski was confronted with a terrible moral dilemma. Should he turn in his brother and risk seeing him sentenced to death, or stay quiet and risk becoming an accomplice to the possible murder of new victims?

After lengthy reflection, he chose to speak to the FBI. This led to the arrest of Ted Kaczynski on April 3, 1996.

"A crude approximation to the truth"

At the start of 1996, when the investigators were confronted with this unexpected lead, they sought to evaluate its credibility. Something troubled them. Ted Kaczynski's biography didn't correspond to the image they had of the Unabomber. Certain elements matched, notably the survivalist lifestyle, but his degree of scholarly achievement and his past as a mathematician were a surprise.

To address the uncertainty, the FBI decided to secretly consult Bill Thurston, who was at the time director of the Mathematical Science Research Institute at Berkeley.

After he read the manifesto, Thurston had no doubts. It was immediately clear to him that it had been written by a mathematician.

I don't know which elements Thurston relied on to reach this conclusion. Redoing the exercise myself (although it's much easier once you know the end of the story), I was above all struck by the last paragraphs. After having proclaimed certitudes for dozens of pages, Kaczynski suddenly points to the fragility of the delirious edifice he had spent twenty-five years constructing:

> FINAL NOTE
> 231. Throughout this article we've made imprecise statements . . . and some of our statements may be flatly false. . . . And of course in a discussion of this kind one must rely heavily on intuitive judgment, and that can sometimes be wrong. So we don't claim that this article expresses more than a crude approximation to the truth.

The concept of truth is at the center of Kaczynski's preoccupations. He blames modern philosophers for having led an "attack against truth and reality."

But what is truth? How could Kaczynski say that his manifesto was only a crude approximation, and yet find it reasonable to kill in the name of this truth? Did he want to suggest that he personally had access to the truth, even though he didn't entirely set it down in his manifesto?

Despite all the while recognizing the weakness of his argument, Kaczynski seemed convinced he was right and the rest of the world was wrong. He was never able to entirely prove it, but that didn't worry him: for him, everything was clear. He had *fabricated* his own truths and certainties, and organized them into a coherent system in which other people and other opinions had no place.

When he learned in the press that, for the first time, one of his bombs had killed someone and that the victim had been torn to pieces, he wrote a celebratory entry in his journal: "Excellent. Humane way to eliminate somebody. He probably never felt a thing."

There's a troubling impression that arises from all this. What if Kaczynski had used techniques of mathematicians to arrive at this frightening level of self-radicalization?

It shouldn't come as a surprise that techniques that seek to reprogram your intuition could turn out to be dangerous. After all, misusing a kitchen knife or even a potato peeler can already send you to the emergency room.

According to the psychiatrist who examined him before his trial, Ted Kaczynski suffered from paranoid schizophrenia. According to Kaczynski, this diagnosis was political persecution. He believed himself to be of sound mind. His lawyers wanted to enter an insanity defense but Kaczynski refused, instead choosing to plead guilty.

Paranoid delirium is a close relation to mathematical reasoning. In a way it's the evil twin. Some people even have difficulty telling them apart. There is, however, an easy way to distinguish them, which we'll get back to.

"What's a mathematician to do?"

In 2010, two years before his death, Bill Thurston still seemed preoccupied with the tragic fate of Ted Kaczynski.

He took the time to write a long response to a question asked on MathOverflow, a collaborative site for the math community. The question came from a user named Muad, apparently a young student lacking in self-assurance. It's entitled "What's a mathematician to do?"

Muad asks how he can contribute to mathematics. He has the feeling that "mathematics is made by people like Gauss or Euler," whose work you can try to understand without this understanding leading to anything new. His fear is that there's nothing new to discover for people like him, normal people, those who don't have any "special talent."

Most math students experience similar feelings at one point or another. Thurston's response offers a radical change of perspective:

> The product of mathematics is clarity and understanding. Not theorems, by themselves.
>
> The world does not suffer from an oversupply of clarity and understanding (to put it mildly).
>
> The real satisfaction from mathematics is in learning from others and sharing with others. All of us have clear understanding of a few things and murky concepts of many more. There is no way to run out of ideas in need of clarification.

Thurston defines mathematics as a collaborative human project oriented toward sharing and understanding, not a search for eternal truths. Without human understanding, theorems have no value. Who cares who proved this or that result first? What counts is the meaning that we give to those results. Real math is the one that lives in each of us.

Thurston's response might seem innocuous, but it's a profound questioning of the way that math has been presented for over two millennia. It's one of the key messages of this book and we'll come back to it.

He continues:

> We are deeply social and deeply instinctual animals, so much that our well-being depends on many things we do that are hard to explain in an intellectual way.

> Bare reason is likely to lead you astray. None of us are smart and wise enough to figure it out intellectually.

It's at this precise point, with the word *astray,* that Thurston includes a direct reference to Ted Kaczynski by inserting a link to his Wikipedia page.

Saying that "bare reason is likely to lead you astray" seems banal, commonsense advice but too vague and unoriginal to attach much importance to it. That's not the case: Thurston is dead serious. He knows exactly what he's talking about and he's making a precise statement.

In chapter 14 we talked about Descartes's project to reconstruct all of science and philosophy from the ground up, relying on our innate ability to find evident that which is true. We said that this approach, *rationalism*, encountered difficulties that Descartes hadn't foreseen.

When Thurston says that "none of us are smart and wise enough" and that "bare reason is likely to lead you astray," this is what he's talking about.

Why is mathematical thought so powerful? What, however, are its limits, and what are the limits of rationality? How do we distinguish between mathematical reasoning and paranoid delirium? The answer to these questions isn't found in mathematics itself but in its close relationship with our language and the inner workings of our intelligence.

This is the subject that will occupy us to the end of the book.

18

The Elephant in the Room

You've always known there's a problem with rationality.

It's supposed to be the basis of our civilization. At any rate that's what they tell you in school. We're taught to organize our ideas in a logical and structured manner. We're taught to distinguish between a reasoning that's valid and one that's not. We're taught to discount what isn't logical, rigorous, coherent.

Of course, no one's stupid enough to believe this story. We just pretend we do. Once the lesson is over, once the school door's closed, we continue to live our lives as if all that was of no importance.

Believing that one day we can become entirely rational is as naïve as believing that one day we'll stop eating sweet and fatty foods.

The paradox is that it doesn't stop us from having recourse to secret rationality.

When you're sincerely preoccupied with something, when you're in trouble, when you have problems at work or problems at home, you instinctively call on the method used by mathematicians.

At night, in your bed, you try to understand the issue. You mull it over. You replay in your head mental images that you dig up from the depths of your memory and imagination. You play Lego with these images. You try to organize them, fit them together and assemble something meaningful, something that makes sense.

Sometimes you have the feeling you understand it all. Your mental images come together. You reinterpret a past event in a new way. You pick up on a detail, a new element, something that had been right under your nose that you hadn't yet perceived.

Now that you see it, everything makes sense. It's a revelation, a discovery that gets you excited and makes you want to share it with others.

You talk about it with your best friend. But soon enough you notice something in her eyes that troubles you. She seems annoyed. She's worried about you. All that she can find to say is a simple phrase: "Try not to rationalize too much."

The worst thing is, you know she's right. You yourself, when someone comes up with a reasoning where everything all fits together too neatly, suspect that something isn't right. He'd thought about it too much, and it seems fishy.

For example, a guy who's spent twenty years in a log cabin, thinking through the root cause of every problem with our civilization, and comes up with a manifesto of 232 numbered paragraphs where everything fits together all too well—you don't find that entirely reassuring. You don't tell yourself, "The guy must be right." Instead you say, "The guy can't have many friends."

If your distrust of rationality was only a matter of intellectual laziness, of not trying hard enough, that would be no big deal. You could let others do the work and benefit from their wisdom.

The problem is that you don't have confidence in the outputs of rationality. You know that thinking and reasoning don't always uncover the truth. Sometimes you have the opposite impression: in some cases, rationality leads you astray from the truth.

It's no small problem. It's an enormous problem. It's the elephant in the room—a problem so enormous and laden with consequence, so central to our existence, that it's never spoken of.

If humanity wants to give itself the slightest chance to overcome the formidable challenges that it is facing, wouldn't it be better if we could start by agreeing whether or not Descartes's method actually works?

Sand and Mud

When Descartes formulated the project to reconstruct science and philosophy from the ground up, it's easy to see where he was coming from.

He remarked that the greatest scholars were incapable of agreeing on the most elementary subjects. Quite often their so-called knowledge consisted only of "magnificent palaces built on nothing more than sand and mud." Mathematics, on the contrary, was built on solid rock. This is what caught Descartes's eye: "I was astonished that nothing more exalted had been built on such sure and solid foundations."

Since the methods used by mathematicians are so effective, since they produce truths that survive millennia without showing their age, couldn't we apply them outside of mathematics and produce unshakeable truths?

We now know the answer, and it is, unfortunately, no, we cannot. Or rather, the answer is partially positive and partially negative.

You can apply the methods used by mathematicians outside of math. I wouldn't have written this book if I didn't want to encourage you to do so. Descartes was right, his method is an incredible tool for understanding the world, and it can literally make us more intelligent. At any rate, there is no plan B. We have no alternate method at our disposal that offers similar benefits.

But when we use it outside of mathematics, we need to be careful: it's only within mathematics that this method is able to produce unshakeable truths.

That doesn't mean you should stop thinking. Being careful doesn't mean opting for confusion and indecisiveness, and refusing to "proceed with confidence through life." Quite the contrary. Rationalizing, looking for explanations for things you don't understand, is a great thing to do. Why would anyone choose to remain ignorant?

Being careful simply means keeping in mind that Descartes's method has the effect of modifying your mental representations and your intuitions and gradually reinforcing their internal consistency.

It's in fact the whole point of the approach. By anchoring our convictions in indisputable evidence and rigorous deduction, we can turn them into certitudes that, over time, become as strong as reinforced concrete.

Except that sometimes these certitudes are false.

A Sort of Unbreachable Wall

You're absolutely certain that all chickens come from eggs, which are laid by other chickens. Through perfectly rigorous reasoning, from a logical perspective, you should thus deduce with absolute certainty that there was never a first chicken nor a first egg, that chickens and eggs have existed since the dawn of time. Chickens and eggs thus preexisted the formation of the planet.

This example is idiotic and that's why it's so important. If such a simple reasoning leads to a conclusion so profoundly flawed, what confidence can we have in reasonings that aren't even idiotic but, instead, are contrived and seem intelligent?

The riddle of the chicken and the egg is supposed to present us with a "paradox," a sort of unbreachable wall to human comprehension, before which we have no other choice but to bow down.

But there aren't any more paradoxes than there are tricks, or truths that are counterintuitive by nature. Being a paradox is always a temporary status, in wait of a resolution. Presenting a problem as structurally being a paradox is just a pompous way of saying you can't solve it.

A superficial resolution of the riddle of the chicken and the egg calls in the theory of evolution. The mother of the chicken we see is another chicken, slightly different from the first. The mother of this second chicken is a bit more different. Everything's fine up to this point. We're still talking about chickens and eggs that look like chickens and eggs.

But if you go back 150 million years, you come across the mother of the mother . . . of the mother of the chicken, who doesn't look like a chicken at all, but a dinosaur. Go back even further, and we come across animals that don't even lay eggs. That might not tell us what they look like but at least it reassures us that chickens didn't preexist the formation of the planet.

This way of solving the riddle, however, leaves aside its most troubling aspect: why was there a riddle in the first place? How can it be that, starting from a hypothesis that is indisputably true, following a reasoning that is indisputably correct, we arrive at a conclusion that is indisputably false?

That's the true riddle of the chicken and the egg, and its solution has been known for almost a century. But this solution is so staggering, unsettling, and consequential that it's carefully obscured and left out of schools.

Here is the reason why there was a riddle: human language is structurally incompatible with logical reasoning, and we can never have 100 percent certainty in truths expressed in human language and arrived at through deductive logic.

That goes for all kinds of "truths," the ones coming from official science and the ones coming from our small everyday reasoning that we employ all the time. And it doesn't matter whether we express our reasoning in words, following structured and logically coherent arguments, or whether we secretly manipulate mental images in our head. It doesn't matter whether we're aware of thinking something through, or whether we let ourselves be guided by our intuition.

The sad reality is that contrary to what Descartes claimed, the "things that we conceive of very clearly and distinctly" are not always true.

Descartes missed an essential point: all reasoning, even the most solid, ends up coming apart the further it gets from day-to-day experience, not for lack of rigor but because our language itself is built on a base of sand and mud.

The only exception is mathematical reasoning when it is articulated in the official language of mathematics. If this artificial language is so inhuman and so incompatible with our usual way of thinking, it's for a very simple reason: its bias is to be compatible with logical reasoning.

When we want to venture far from our everyday concrete experience, logical formalism helps guide us. It's the only tool at our disposal that lets us give free rein to our impulse toward rationality without limits, without complexes or taboos.

Outside of mathematics, rationality remains under constant threat from the fragility of our language and our way of perceiving the world.

Elephants Don't Exist

In chapter 6, we already stated that the official language of mathematics, logical formalism, is a foreign language for humans, and we gave this example to show how it works: in mathematics, if having a trunk is part of the definition of an elephant, then an elephant without a trunk immediately ceases being an elephant.

This bizarre, perversely rigid relationship to the meaning of words is at the heart of the mathematical approach. It's an oddity that often shocks beginners. But there's a simple explanation for it: without this rigidity, no language can be compatible with logical reasoning. It may be annoying, but that's how it is.

Any reasoning about elephants that makes use of the fact that they have trunks will be invalidated by the discovery of a single elephant without a trunk.

In chapter 8, we touched upon a related topic when discussing the shortcomings of dictionaries. At a distance, dictionaries look like the real thing. We always want to believe that the words we use are solidly defined and that the phrases we speak have a precise meaning. But once you scratch a bit below the surface, you find circular definitions.

The error would be in believing that people who put together dictionaries aren't doing their job correctly, and that there's a better, smarter, and more rigorous way of defining the words we use.

But there's a deep structural reason why dictionary definitions are so deficient: it's rigorously impossible to truly define words in our language, and our relationship to the world is much less solid than we would like to believe.

Starting with a word from everyday language, trying to solidify the definition, and finding out that it's an impossible task: it's a troubling yet instructive experience that you should take the time to try at least once in your life.

Let's do that with elephants. I found this definition in a dictionary: "a thickset, usually extremely large, nearly hairless, herbivorous mammal that has a snout elongated into a muscular trunk and two incisors in the upper jaw developed especially in the male into long ivory tusks."

The approach has the merit of being pragmatic. It consists of listing the different characteristics you can expect to find in an elephant.

It is nevertheless a circular definition: how do you think they define a "trunk"? or a "tusk"? or "ivory"?

Apart from circularity, this definition also has the flaw of being silent on an essential aspect, the concept of animal species.

It's obvious to all that we're not dealing with animals that are isolated individuals, but with beings that we can group into species. If we've invented this word *elephant,* it's because we had the impression that the individual elephants in front of our eyes all shared something in common. By "elephant," it's this common thing between them that we're referring to, the *species.*

In reality, as you know, elephants make up not one but two species, African elephants and Asian elephants.

In fact, it's more complicated than that. For more than twenty years biologists have known that there are two distinct species of African elephants: the savanna elephant, whose scientific name is *Loxodonta africana,* and the forest elephant, *Loxodonta cyclotis.* Asian elephants form a third species, *Elephas maximus.*

If you really wanted to give a noncircular and seriously scientific definition of the word *elephant,* you might come up with something like "Generic name given to representatives of three species, *Loxodonta africana, Loxodonta cyclotis,* and *Elephas maximus.*"

Except that now you'd have to give a serious definition of *Loxodonta africana, Loxodonta cyclotis,* and *Elephas maximus.* And as shocking as it may seem, no one is able to do so.

In practice, biologists define species as starting from a given individual, called a *holotype,* that serves as a point of reference. For example, there's a specific elephant (who's been dead for a long time) that has the honor of serving as the holotype for *Loxodonta africana.* It is "elephant zero," the standard-bearer of the species. From a scientific point of view, *Loxodonta africana* is nothing other than the group of individuals, living or dead, "of the same species as" this elephant zero.

It simply remains to give a precise meaning to "of the same species as." And here's where things get tricky.

In any reasonable definition of an animal species, you'd want to be able to say that a mother is of the same species as its children. But if you take on one hand the mother of the mother . . . of the mother of elephant zero, and on the other hand the mother of the mother . . . of your own mother, at a certain point you'll get to the same female, belonging to a species of extinct mammals that lived more than 150 million years ago, alongside dinosaurs. It's the exact same problem of the chicken and the egg. The logical conclusion is that you are an elephant.

To avoid this problem, it's necessary to set a limit: you're of the same species as your mother, and the mother of your mother, and so on, but you can't continue that indefinitely.

How do you determine the limit? The official solution relies on the concept of *interfertility* and leads to the following definition, which you can find on Wikipedia under "species": "In biology, a species is often defined as the largest group of organisms in which any two individuals of the appropriate sexes or mating types can produce fertile offspring, typically by sexual reproduction."

The issue of fertile offspring notably concerns donkeys and horses, which can produce offspring (mules) that are, however, sterile. They are thus two different species.

You might think that we've finally nailed it. But that's far from being the case. According to this definition, a sterile individual is a species unto itself. When you neuter your cat, it changes species. That obviously makes no sense, but it's the result of the strict application of the definition.

More seriously, what do you do when the offspring are generally but not always sterile? There are a number of documented examples of fertile mules. Where do you set the limit? Is it necessary to fix a threshold, arbitrarily decree that less than 1 percent of a chance of producing fertile offspring means you're dealing with two distinct species?

The notion of interfertility is intrinsically vague and problematic. When two species separate, at the precise moment when the definition would be of most value, it ceases to function. This concerns the origins of our own species: our DNA carries traces of fertile hybridizations with Neanderthals. Does this mean that *Homo sapiens* and *Homo neanderthalensis* form one and the same species, contrary to what is communally admitted?

Apart from these theoretic inconsistencies, the biological species concept creates immense practical issues. If I want to know whether I'm a savanna elephant, I should mate with a female *Loxodonta africana.* What do I do if she's not in the mood? Or if she doesn't get pregnant because it's not the right moment? How many times do I need to try?

Biologists are supposed to have a more developed practical sense than mathematicians. I'd be curious to know how they do it.

Vaguely Knowing What We Mean

The strangest thing about all this is the contrast between our inability to give a 100 percent rigorous definition of an elephant, and the evidence with which the concept presents itself to our intuition.

When we hear the word *elephant,* we have the impression that we know very well what it means. When we encounter an elephant, we recognize it right away. We have a perfectly clear notion of what elephants are.

The problems come along only once we try to specify the slight vagueness that surrounds the first definition we come up with, when we want to make it compatible with logical reasoning. Our attempts to specify our thinking always end in incoherence, creating inconsistencies that we have to resolve each time with new scientific discoveries, which in turn lead to new inconsistencies.

Something simple, concrete, and plainly *obvious* seems impossible to really capture in words.

This oddity isn't specific to the notion of an elephant. It's a universal phenomenon that reflects the neurological underpinnings of our mental processes. We'll come back to this in the next chapter.

The most spectacular illustration of our inability to give a precise meaning to words comes from the pen of Charles Darwin himself, in 1859, in the opening lines of the second chapter of *Origin of Species:* "Nor shall I here discuss the various definitions which have been given of the term species. No one definition has as yet satisfied all naturalists; yet every naturalist knows vaguely what he means when he speaks of a species."

It is quite telling to find such a blatant admission of weakness at the heart of a timeless scientific masterpiece. The issue runs deep and there's not much we can do about it.

Darwin knew only vaguely what he meant by species, and yet he wrote an entire book about it.

Two Languages, Two Sets of Rules

In the end, if you really want to understand what math is and why it is omnipresent in science, you have to start with this fragility of our language.

Mathematics is as old as our desire for reason, and, to this end, to secure and stabilize the meaning of words. It's a common error to restrict it to the study of numbers and shapes. Beyond its technical aspects, beyond theorems and equations, math is above all a different way of using language and attributing meaning to words.

Human language and mathematical language have evolved in parallel over millennia. Today they have become so intertwined that it's become hard to tell one from the other. In day-to-day life, in the most ordinary kinds of conversations, we navigate between these two ways of using words, generally without being aware of it.

That goes for all of us, even people who think they hate math. Whatever our background and academic level, we have all acquired a certain degree of familiarity with the mathematical approach and have daily recourse to the modes of thinking it enables.

And since no one's ever explained to us how the whole thing works, we keep getting tripped up. We pass from one language to the next, forgetting that they follow two totally different logics. In some cases, this leads to serious consequences.

Each of these languages has its own functions, its own set of

Two languages, two sets of rules

	Human language	*Mathematical language*
Means of defining words	Shared perceptions	Axiomatic characterizations
Strengths	Direct relation to reality, the meaning of words is self evident	Coherence, precision, stability of meaning, one can unambiguously speak of invisible things
Weaknesses	Vagueness, incoherence, unstable meanings	Not human, impossible to intuitively interpret 100% correctly
Compatible with logical reasoning	No	Yes
Outcomes of rational thinking	Explanatory hypotheses, theories, predictions	Theorems
Means of verification	Confrontation with reality	Logical proof

rules, its own strengths, its own weaknesses. And they're equally indispensable.

Taking Words at Face Value

The two languages often use the same words. What changes is how we attribute meaning to those words.

The word *sphere* is a good example. When you hear someone say, "The Earth is shaped like a sphere," the meaning of these words is pretty clear to you.

If, like me, you think the statement is correct, it is because you interpret the word *sphere* in human language, in a perceptual and vague manner.

In mathematical language, however, the statement is clearly false: spheres can't have mountains.

In mathematics, words are defined "axiomatically": via formal definitions that characterize them entirely. They are imaginary, perfect, and fixed constructions: a sphere is "the set of all points in three-dimensional space that are located at an equal distance from a center." You can't change any of this. If you take out or even slightly shift a single point, it ceases to be a sphere.

The worst thing is that "sphere" has the exact same definition in human-language dictionaries. The only thing that changes is our relationship to the definition.

In human language, no one ever really takes words at face value. In real life, what you call a sphere corresponds to what you perceive of as a sphere. To decree that a given shape is a sphere, you have a tolerance margin. An orange is a sphere, an apple is more or less a sphere, a pear isn't a sphere. I challenge you to write down the exact contours of the tolerance margin that is implicit in your perceptual definition of a sphere.

Rationalism vs. Empiricism

More interestingly, you can do it the other way. You can start with a word in human language and pretend to treat it as a word in mathematical language. It's what we all do whenever we reason.

It's a bit tricky, but we're experts at doing it without ever noticing. We start with human language, shift to mathematical language for reasoning, and return to human language. We do this each time we formulate hypotheses and try to draw conclusions from them.

This day-to-day activity is an instance of what is pompously called *the scientific approach.* I'll sum it up as follows through an example that is rather simplistic but faithfully illustrates all the steps in the process.

If you state, for example, that by definition an elephant is "a representative of one of three species, *Loxodonta africana, Loxodonta cyclotis,* and *Elephas maximus,*" if you make believe that this definition states an absolute truth, then it is possible to draw logical conclusions from it.

That's what dictionary definitions are for: they serve as *mathematical models* that temporarily anchor the meaning of words, giving us a chance to reason with them.

When you have an elephant in front of you that is neither *Loxodonta africana* nor *Elephas maximus,* you can deduce that it's *Loxodonta cyclotis.* Within your model, this deduction is 100 percent reliable. You're absolutely sure: you "did the math."

But is this deduction correct in real life? It all depends on the reliability of your model. Perhaps you're out of luck (or incredibly lucky) and you have a fourth species of elephant before you, so rare that it hasn't yet been described.

In human language, nothing is ever 100 percent reliable. We're constantly being surprised. It's why a scientific theory only ever makes predictions, which can then be validated by experience (and the theory gains credibility) or disproven (in which case you need to change the model).

Models are not good or bad in and of themselves. Saying that the Earth is shaped like a sphere is a good enough model as long as you don't try to deduce from it that mountains don't exist. The success of our technology is proof that the scientific approach works: even if it doesn't produce *absolute* truths, science gives us a powerful method of thinking and its predictions are close enough to reality to be of practical use.

In the end, rationality should be used as a guide rather than an ultimate judge. The reality that's before our eyes always merits more attention than the certitudes in our heads. Rationality is great, but empiricists do have a point.

Trusting reason too much, using human language as if it had all the attributes of mathematical language, as if words had a precise meaning, as if each detail merited being interpreted and the logical validity of an argument sufficed to guarantee the validity of its conclusions, is a characteristic symptom of paranoia. When applied outside of mathematics and without any safeguards, mathematical reasoning becomes an actual illness.

A Torn Spider's Web

Our ancestral addiction to overthinking is without a doubt the source of terrible misunderstandings that surround the notion of truth.

When I speak of "truth," I'm speaking of the truth of mathematicians, the absolute and eternal truth, what some people like to write as Truth, or TRUTH, or sometimes even **TRUTH.**

This sort of truth is a mathematical concept. It exists in the same way as the number 5 or triangles and rectangles. It is undoubtedly the first invention in the history of mathematics, that which preceded all the others and which has had the greatest impact on our culture.

Mathematical "truth" has its counterpart in human language, just like "sphere." But the human version is very problematic. Like a fruit that doesn't tolerate shipping well, the concept of truth suffers in translation. A damaged sphere still looks like a sphere, but a damaged truth doesn't look like anything at all.

Besides, we never expect statements in human language to be definitively and implacably "true." We simply expect them to be clear, expressive, interesting, honest, sincere, and able to teach us something useful and relevant about the world.

When we say that something is "true," we never mean it literally. We only ever use the word as a shortcut for all these other things, because otherwise we'd have no occasion to use it.

This situation is of course frustrating. We'd like the world to be clearer and more stable. We'd like truth to be more solid and less dependent on our point of view.

The Austrian philosopher Ludwig Wittgenstein (1889–1951) perfectly summed up this frustration: "The more closely we examine actual language, the greater becomes the conflict between it and our requirement."

Logic doesn't function unless words have a definition that is explicit, perfectly precise, and stable over time. Despite immense efforts, we've been unable to produce these kinds of definitions outside of mathematics. Wittgenstein affirms that it's a quixotic quest: **"We feel as if we had to repair a torn spider's web with our fingers."**

In acknowledging the intrinsic limitations of our language, Wittgenstein made one of the great philosophical breakthroughs of the twentieth century. This allowed him to break with a multi-millennial tradition dominated by metaphysics, in which philosophers believed that it was possible to attack, using rationality, problems that were strikingly similar to that of the chicken and the egg: problems so remote from our daily experience that they were occurring only as a result of our language losing its grip.

It's surprising that Wittgenstein's philosophy isn't more well known outside of specialized circles, as it holds a very practical life lesson: we should accept going step by step, clarifying our language as we move forward, and being regularly surprised. It's an effective antidote for paranoia.

Another key lesson we should have drawn from him concerns the way we teach mathematics. The lesson is very much in line with Thurston's remark that "The product of mathematics is clarity and understanding. Not theorems, by themselves."

If math were good only for producing eternal truths, it would be of strictly no use, since there is no place for eternal truths in the human experience (as our language simply doesn't permit it).

Yet we continue to teach math. This indicates that we continue to believe that math is useful in a certain way. So what is it that math is good for?

To find out, to understand how math really works and what it can really do for us, we cannot continue to overlook its most direct practical aspect: math works on our brain and modifies how we see the world.

19

Abstract and Vague

Are you familiar with this optical illusion? It's one of the oldest and most famous.

What do you see?

Look closely.

Most people will say they see an elephant. I imagine you see one as well. But can you see something else?

Take your time.

Give yourself a real chance before turning the page and reading what follows.

The most extraordinary thing about this illusion is that it takes quite a bit of effort to make it go away. It's nearly impossible not to see an elephant. And yet, if you look closely, you'll notice that there is no elephant, just ink on a page.

Between ink on page and an elephant, the difference is indisputably massive. What mysterious phenomenon makes us see an elephant where there's only ink on a page?

This type of optical illusion is nowadays called a *drawing*. A drawing, of course, is a lot more than an illusion. A drawing has style and artistic value, it carries a message, a symbolic and cultural significance that can't be reduced to its figurative content.

And yet, if we instantly recognize mammoths painted on cave walls by Paleolithic people whose language, customs, and beliefs we know nothing of, it's because something in the drawing goes beyond any cultural convention. Even if we don't have the codes, we understand it all the same. Our brain automatically makes the connection between a *drawn* animal and a *real* animal. It's in this way that drawings are true optical illusions.

Visual comprehension of drawings develops from infancy and doesn't require any special teaching. Babies understand drawings before being able to talk. They understand them so well that we rely on drawings to teach vocabulary. If drawings were only cultural conventions, it would be the other way around: babies would begin by learning vocabulary and then be able to learn how to recognize drawings.

A number of animal species (and not just mammals) are known to be capable of recognizing drawn objects without having been taught. Understanding drawings isn't a strictly human privilege.

The illusion at work in a drawing is something really extraordinary.

Just consider the amount of information that you managed to gather from the little sketch that opened this chapter. Is the elephant young or old? Dangerous or harmless? Is it angry? Does it have a lot of self-confidence? Do you feel sympathy or distrust?

You've never taken a class on interpreting drawings and yet you know how to answer these questions, immediately and without any effort. All that from a few lines drawn on a page.

How is such a miracle even biologically possible?

The explanation that current science is able to provide will help clarify the nature of our learning process, and how the mental plasticity described since the beginning of this book actually operates.

The Mystery of Vision

Vision is a complex phenomenon that concerns optics, biochemistry, and neurology. The organs of vision comprise not only the eye but also the optic nerve and above all the brain.

Knowing what eyes are used for is simple enough: roughly speaking, they're cameras. The metaphor is simple, and even a bit simplistic, but it's relevant and reasonably correct. There are many unresolved scientific questions about how our eyes work, their development in the embryonic stage, and the evolutionary processes that led to their emergence. These are legitimate and difficult questions. But we know enough to have reached the point where it'd be ridiculous to claim that eyes are something of a mystery.

As for the optic nerve, it can be seen as a kind of cable linking the eye to the brain. Here again the metaphor works quite well (although the optic nerve does also perform some signal preprocessing).

On the other hand, what happens once you get past the optic nerve and inside our visual cortex is a very intriguing phenomenon that has long resisted our efforts to understand it. You can legitimately say that it has long been one of the great mysteries in the history of science.

With a camera, an image is formed of pixels: it's a grid in which each box has a certain value of red, green, and blue luminosity. The mystery concerns the way our brain treats the raw information received from the optic nerve to "extract meaning" and to "recognize" what is in the image.

Here is a good way to sum up the problem: by knowing the luminosity and the color values of all the pixels, how is it that you know there's an elephant somewhere in the image?

The Concept of "Elephantness"

The ease with which we recognize elephants is all the more troubling since, as we've seen in the previous chapter, we're unable to really define what they are.

This is no coincidence. We'd like to define elephants as we perceive them, because that definition would have the most sense for us. But the method that our brain uses to recognize elephants is at the same time stunningly efficient and perfectly impossible to translate into words.

In fact, it's so efficient that it's hardly believable.

First off, your ability to recognize an elephant doesn't depend on the angle you're looking from. Whether you see it from in front or from behind, from the side or in three-quarters view, standing upright or lying down, whether it's big or small, however it's moving in relation to your own movement, you recognize it instantly. You also have an incredible tolerance for a multitude of abnormalities and unexpected characteristics an elephant might exhibit. Whether the elephant is albino or painted with geometric shapes or colored stripes, you'll still know it's an elephant.

From a strictly visual point of view, however, that is, from a point of view of the luminosity and color of the pixels in the raw image, these anomalous situations really don't have much in common.

But that's not all. When a child sees a real elephant for the first time, even if they've never seen a picture or heard it spoken of before, if you point your finger at the elephant and say, "That's an elephant," the child knows immediately what you're talking about.

That's not as obvious as it may seem. What keeps the child from thinking that what you're calling an *elephant* is simply the left front foot, or the trunk, or a piece of the trunk, or a fly sitting on the trunk?

If the child understands right away, it's because they already see the elephant. They noticed it immediately, well before you said what it was. The elephant stood out as something remarkable that deserved a name. They were probably getting ready to ask you what it was.

Without this ability, our language simply wouldn't exist. We wouldn't be able to explain what words referred to.

There's something even more surprising. If, instead of meeting elephants for the first time in the flesh, your child began by seeing grotesquely cartoonish drawings of them, that wouldn't be a problem. They'd be perfectly convinced they knew what an elephant was. The day they saw one *for real* they might be surprised and probably intimidated by its size, but they'd easily recognize the animal they already knew.

The conclusion to all of this is that our brain seems to automatically extract, from the raw visual data fed continuously into it by the optic nerve, a universal idea of what an elephant is. It then becomes able to recognize this abstract concept of an elephant through its multiple incarnations, in situations so remarkably varied it would be ridiculous to try to list them all.

Without trying, as if by magic, we develop a curiously reliable sense of "elephantness" by mere exposure to scenes involving elephants.

At the start of this process, the elephant is nothing but a strange impression mixing the familiar and the bizarre. We see that it's an animal, with a surprisingly large nose, legs like tree trunks, and ears like giant fans. But this animal doesn't resemble any we know. It greatly intrigues us and we feel it deserves a name.

The concept of elephant first emerges in our visual cortex in the form of this impression. With repeated observations, the image stabilizes and becomes clearer. At the end of the process, elephants become so natural and familiar to us that it's as if we've always known what they are.

But what is a concept? Why do we think with concepts? At what level of reality do they exist? What stuff are they made of? What makes us able to perceive them?

These questions are among the most ancient in the history of philosophy and, for millennia, they have been said to be insoluble. Asking these questions, of course, is asking how our brains work.

Interestingly, the mystery of vision reframes these questions without any need for abstract or confusing language. It's a very practical, very down-to-earth way of asking questions about the inner workings of our intelligence.

The Worst Metaphor

Our brain is often compared to a computer. This metaphor is correct in two aspects: both the brain and computers are capable of accomplishing complex tasks of information processing, and they both make use of electrical signals.

As for all the rest, the metaphor is catastrophically false. It ruins our chance of understanding what is going on.

A computer is a perfect embodiment of System 2: it's a machine capable of mechanically applying long sequences of logical instructions at breathtaking speeds without making mistakes—something our brain is entirely unable to accomplish.

A computer is made of a central processing unit where calculations are made and memory units where information is stored. Between these distinct units, information circulates at high speeds along electrical circuits without being transformed. In our brains, it's quite the contrary. Information circulates slowly and is transformed along each step of its circulation. Memory, processing, and circulation are indissoluble.

A computer strings instructions one after the other, sequentially, paced by an internal clock that ticks billions of times per second. The time it takes to activate a connection between neurons is on the order of a thousandth of a second. The base operations of our brain are thus *a million times slower* than that of a computer. But our brain isn't sequential: it processes in parallel *billions and billions* of these operations.

The silicon circuits of computers are immutable, engraved in an inert material. Our brain is living tissue that constantly reconfigures itself.

A Perceptual System

Rather than representing the brain as a system for doing calculations, it's much more illuminating to see it as a perceptual system. Our brain is the central organ through which we perceive the world. It allows us to sense things: for example, that an elephant is in front of us.

Each neuron in our brain is in itself a tiny perceptual system. Anatomically, a typical neuron consists of three parts:

—A treelike structure that branches out in thousands of small receptors called *dendrites.* The dendrites are the receiving end of the neuron.

—A central part called the *soma:* this is the body of the neuron and contains the nucleus.

—A kind of trunk or stem called the *axon* that branches out and ends in what are called *axon terminals.* This is the transmitter end of the neuron, which communicates with other neurons.

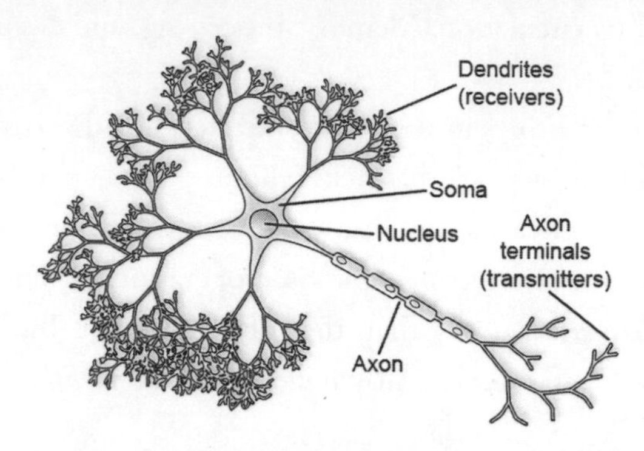

Neurons are connected to one another in a very specific orientation: the dendrites of one neuron are plugged into the axon terminals of other neurons, enabling it to collect information from them. The connections thus formed are called *synapses.*

Neurons exhibit an all-or-nothing type of behavior: they can either be in a resting state or suddenly "fire up" in full force, in which case an electrical *action potential* travels down their axon to the terminals, triggering the release into the synapses of molecules called *neurotransmitters.*

These neurotransmitters are in turn detected by the dendrites of receiving neurons.

To decide whether it should fire up, a neuron essentially conducts a poll. If enough of its dendrites detect that upstream neurons have just fired up, the neuron will itself fire up, which in turn may trigger the firing up of downstream neurons.

A neuron is a perceptual system with a narrow worldview and a binary response: all it perceives from the world is the activity of the neurons that are immediately upstream and, apart from resting, all it can do is fire up.

An Emergent Property

The first time someone tried to explain to me how neurons worked, it didn't interest me at all. It didn't seem to lead anywhere. If our neurons are so primitive, how can we be intelligent?

The mechanisms of our intelligence are impossible to understand as long as you try to locate them in a specific place in our brain. Intelligence is what is called an *emergent property:* individually our neurons are primitive and limited, but vast assemblies of neurons make incredibly sophisticated behaviors "emerge" that can't be attributed to any neuron by itself—these large-scale behaviors are what we call *intelligence.*

It's a bit like traffic jams: you can spend twenty years of your life reverse-engineering cars, but that won't teach you anything about traffic jams. And yet traffic jams exist and they're entirely made up of cars.

An enormous divide separates the individual behavior of our neurons and the overall functioning of our brain. For a long time, this divide seemed so enormous that scientists despaired of ever understanding it.

That's no longer the case. The mystery of vision is now largely resolved. We don't understand everything, but what we understand is sufficiently detailed and makes enough sense that, to some people, it no longer feels like there is a mystery (although, of course, many deep questions remain open).

First, thanks to progress in neurology, we now have a much better understanding of the overall organization of our brain and the wiring diagram of our neurons. It has also become possible to follow in real time the activity of individual neurons or specific regions of the brain in both humans and animals.

This, in itself, wouldn't solve the mystery. Brain-imaging technologies are still incapable of mapping out the entire brain activity with enough resolution to fully understand what's going on. We're still very far from being able to simultaneously follow, for example, all the neurons at work in recognizing an elephant—and even further from being able to follow these neurons throughout the lifelong process of learning.

The most spectacular and compelling breakthrough has come from another discipline: computer science. Since the 1950s psychologists and computer scientists have looked for inspiration in the functioning of our neurons and the anatomy of our cerebral cortex to construct systems of *artificial intelligence.* Because they imitate the architecture of our brain, the behavior of these systems sheds light on what's going on in our heads.

Frank Rosenblatt (1928–1971), one of the pioneers of this approach, helped construct the first mathematical model of a neuron and fashioned a computing device that implemented this model. But modeling the behavior of complex neural networks capable of simulating our ability to see was a problem of an entirely different scale. The technology stumbled along for decades and went through numerous ups and downs. At some point, the AI community grew so disillusioned that artificial neural networks were seen as a technological dead end. Three scientists, Geoffrey Hinton, Yann LeCun, and Yoshua Bengio, continued to believe in the approach. History proved them right.

Toward the end of the 2000s their "deep-learning" algorithms had made so much progress that they had become capable of resolving advanced problems in the recognition of images, such as the detection of the presence of elephants.

An Effective Metaphor

Around 2010, when I began to familiarize myself with these algorithms, I was excited to discover a way of describing the process of understanding that, for the first time, was compatible with what I had personally experienced.

From the outset in this book, I've discussed a number of powerful and mysterious phenomena that have perplexed me throughout my mathematical career: mental plasticity, the inevitable ambiguity of human language, the role of time and trial and error in trying to understand things, the necessity to ask stupid questions, the feeling of obviousness that comes after the fact.

With deep learning, the process of understanding could be made tangible and concrete. It had finally become possible to speak of it without invoking some kind of black magic.

The subject fascinated me so much that I decided to call an end to my career in mathematics. I had just completed an important cycle in my research in algebra and geometry, and I saw an opportunity to explore a radically new theme, one that might be able to shed light on what I'd experienced.

I chose to approach it in the most practical manner possible, by quitting my academic position and founding an artificial intelligence startup.

To explain the nature of our intelligence and the mechanisms of our thought, deep learning offers the best metaphor I know of.

An Elephant Neuron

The first mystery that deep learning allows you to dissipate is that of the emergence of concepts. In other words, what had been for millennia one of the liveliest debates in metaphysics was suddenly reincarnated in the realm of software, confronting us with an undisputable experimental reality: conceptual thought spontaneously emerges in vast assemblies of artificial neurons subjected to unstructured data, for example, a flood of images.

Roughly speaking, here's how it works in the context of vision. Deep-learning algorithms model our cortex as a neural network with multiple layers. The first layer is the raw image: a matrix of neurons that represent pixels. The second layer is formed of neurons whose dendrites are linked with neurons in the first layer. The third layer is formed of neurons whose dendrites are linked to the neurons in the second layer, and so on. It's because the network is made up of many superimposed layers that it's called "deep" learning.

In my description of how neurons function, I omitted one important detail: when a neuron runs a poll of its dendrites to decide whether it should fire up, the poll isn't democratic. Each connection in a neural network carries a certain "weight" that determines how much it counts toward the decision.

When the network is subject to a flood of raw images, it gradually adjusts all the weights according to a mechanism I'll explain in a few pages.

It's through this process of adjusting weights that the network "learns" and "becomes intelligent."

When you let a deep-learning algorithm run for a long time, for example, by making it "learn" from millions and millions of photos taken at random from the internet, you notice that each neuron gradually comes to specialize in the detection of a certain "concept."

The concepts of the first layers are very primitive, while those of the deeper layers are much more sophisticated.

For example, a neuron in the second layer might specialize in the detection of a vertical line in the bottom left corner of an image, or in the slight change in luminosity in another region of the image. It will fire up exclusively when this element is present.

In the third layer the concepts become slightly more sophisticated. For example, a neuron might detect certain types of angles between two segments situated in a certain zone of the image.

As you get further into the network, the concepts grow in richness and abstraction. They become more and more "deep." In the fifth layer, certain neurons might, for example, specialize in the detection of triangles or certain types of curves.

In the twentieth layer, a neuron might specialize in the detection of elephants—whether they're real or drawn.

This way of presenting things is purposely simplistic. The reality is more complicated than that, and the correspondence between neurons and concepts isn't necessarily that direct. Specific experiments do suggest that, for each famous actor or actress you know, you really have a specific neuron that reacts specifically to his or her presence on the screen (see "Notes and Further Reading" section). But some scientists think that concepts correspond to groups of neurons rather than individual neurons. In computer simulations, individual neurons sometimes specialize in the detection of high-level objects such as elephants, but in other instances this detection mobilizes a whole group of neurons rather than a specific one.

While simplistic, this model does capture some of the most salient aspects of our cognitive processes. Saying that in your brain there is a dedicated elephant neuron is a bit of a stretch, but it's an illuminating one. This is why we'll proceed as if it really were the case.

One Hundred Trillion Filaments

Your elephant neuron has thousands of dendrites. Your personal definition of an elephant involves thousands of criteria, which are themselves abstract attributes at a fairly high level, such as "is an animal," "has a trunk," "has big ears," "is gray," "is big," "trumpets," "has ivory tusks," "has rough skin," "moves in such and such manner," and so on.

Each of these attributes has its own weight. It's a good bet that "has a trunk" has a high weight, because having a trunk is a highly relevant feature. Your elephant neuron calculates an "elephantness score" by adding up the weights of the attributes whose neurons are activated.

Once the score passes a certain threshold, the neuron decides that you're dealing with an elephant. Beneath this threshold there's a gray zone where you're not sure it's an elephant (and where someone else might have a different opinion), then a zone where it's clearly not an elephant.

It's the large number of criteria at play that makes your elephant-detection system so robust and reliable. Your elephantness score is sufficiently well sampled to stay relevant in unforeseen situations and to tolerate a large variety of anomalies.

The exact definition is evidently impossible to write. An entire book wouldn't be enough, and at any rate you'd never find the words.

This tangle of 100 trillion neural connections is the spider's web Wittgenstein was talking about. It's unthinkable to unravel all of that. But without unraveling all of that, there's no chance of being able to define anything.

Deep-learning algorithms, even the most powerful and sophisticated, are only gross simplifications of our cerebral architecture. Our cortex is indeed structured in layers, though not as strictly and narrowly as in computer models: your "elephant" neuron polls your "trunk" neuron, but your "trunk" neuron itself certainly polls your "elephant" neuron. The circularity of definitions is impossible to avoid.

It is also an oversimplification to imagine that our brains are organized in specialized regions perfectly isolated from each other: vision happens within a broader context and, at any rate, your definition of elephant isn't entirely visual.

The Organic Process of Learning

It remains to describe the process of learning itself: what mechanisms do the neurons rely on to determine the "weight" of their connections with other neurons?

Let's go back to the example of your elephant neuron. It is constantly analyzing the state of its upstream neurons to decide whether or not it should fire up. You're scrutinizing the world in real time, on the watch for elephants. (It's always an abuse of language to speak of "real time," because no system actually functions in real time. It takes a neuron around half a millisecond to fire up.)

In chapter 11, we called this *System 1,* instantaneous intuitive thinking, that which gives you the impression of thinking as fast as lightning.

In parallel to this, another phenomenon takes place in the background. It happens at a much slower pace, and is so discrete that we can't perceive it. The correct metaphor isn't lightning, but organic growth. It's the process through which we learn. It's the basis of what we have called *System 3,* our ability to gradually modify the way we represent the world to ourselves.

If one day you come across an elephant without a trunk, you'd be surprised.

What does it mean "to be surprised"? A trunkless elephant surprises you because your vision of the world hadn't anticipated it. That still doesn't keep you from understanding. You'd almost certainly still see that it's an elephant, while having the disturbing feeling that something's terribly wrong.

When you mathematically model a deep-learning system, you can define a numerical quantity that measures its "perplexity" in a given situation. A system that learns is one that adjusts its weights in order to reduce its perplexity.

Intuitively, this is what perplexity means. In your elephantness score, "having a trunk" carries a pretty high weight, because elephants and trunks usually come together. While the other criteria will allow you to compensate and "see" an elephant despite the absence of a trunk, it's an abnormal situation and you feel it *physically,* whether you're aware of it or not.

Your elephant neuron is perplexed. It detected an elephant, yet there was no trunk, and up to this point trunks were supposed to be an essential feature. The neuron factors in this new reality by slightly diminishing the associated weight. If you continue to come across elephants without trunks, you'd end up by hardly taking this criterion into account at all.

In reality there's no need to have such large abnormalities for your neurons to correct their weights. They're constantly, if slightly, in adjustment with each stimulation. Physiologically, that corresponds to the ability of synaptic connections to strengthen or weaken. New connections are created and others disappear. Our mental circuitry is constantly reconfiguring itself.

Mental plasticity is nothing more than this: the decentralized action of your neurons that, individually, seek to reinforce the consistency of their score.

The most extraordinary thing—and it's been perfectly demonstrated experimentally thanks to deep-learning algorithms—is that such simple mechanisms allow high-level abstract concepts, such as elephants, to gradually emerge, starting from a state of departure where the connections and weights are chosen at random.

You weren't born with an elephant neuron. The first time you saw one, you were greatly perplexed: your "animal" neuron was excited, as well as your "enormous thing that deserves my full attention" neuron, as well as many other neurons corresponding to the many attributes you could recognize. But this powerful and complex impression had no name. You looked at it carefully, to take it in and to learn.

In your head, the elephant was at first a composite object, mobilizing a large number among the 100 billion neurons in your brain. One of the neurons that fired up during this first elephantine encounter had a special destiny. Little by little, by gradually adjusting its weights, it became more and more specialized. Over time, it *became* your elephant neuron.

Concepts emerge in deep-learning networks under the simple effect of exposure to the world. They emerge literally out of nothing, as waves are formed on a flat ocean by the effect of the wind: initially, there are just tiny random irregularities on the surface of the water, but these irregularities are then amplified by feedback mechanisms, following laws of physics that are simple to describe on the microscopic level and that, on a greater scale, give rise to incredibly complex emergent phenomena.

Abstract and Vague

The scientific, technological, and philosophical implications of all this go far beyond the scope of this book. Here's how we can sum it up as far as our needs go.

Our brain, like any animal brain, is a perceptual machine that constantly fabricates abstractions. We construct and we maintain a representation of the material world through the tangled network of our neural connections. This representation of the world is a piling up of layers upon layers of abstractions. Down to its very core, it's conceptual in nature.

Conceptual thought isn't a human privilege. It doesn't arise from our language or our culture. When saying this, I'm using the word *thought* in a very broad sense, to designate the neurological processes that constitute the substrate of our intelligence. Any lion thinks in a conceptual manner, and has an elephant neuron in its head.

The flaws of our language are but a reflection of its neurological underpinnings. The meanings that we assign to words are perceptual: we know how to recognize an elephant but we can never really define what it is.

Every definition is an approximation. The meaning of words is always fluid, ambiguous, changing. Nothing is ever clear-cut. Inside our head, the world is abstract and vague.

20

A Mathematical Awakening

All throughout my mathematical journey, despite my taste for math and the pleasure it brought me, I've always had the impression that the real challenge was elsewhere.

What really mattered, what motivated me and made me want to continue, wasn't the theorems that I could prove and that would interest only a few specialists, but something else. This other thing was much more profound and much more universal.

It even seemed to be incredibly important. Yet I couldn't really explain what it was, and I wasn't even able to give it a name.

This troubled me for many years. I had the weird feeling, which is familiar to many creative mathematicians, that something was going on, something unclear that deserved an explanation. I had no idea what it was, but I knew that it had to do with human understanding. That gave me a rough idea of which direction to follow.

Mathematical research seemed to me to be the best way to approach it. I was like an explorer who set out to discover an unknown continent that is barely sketched out on the map. I had no idea what I'd find, but it was perfectly clear from the start that, in reality, I was seeking to discover myself.

This book is the story of my adventure. I lived through it so I could tell the tale.

Years have passed and I'm finally able to put into words the thing that, at the time, seemed so strange and unclear to me. It's the subject of this final chapter.

Fallen from the Sky

When I was a PhD student and someone asked me about the usefulness of my research, I sidestepped the question with a joke: "It will be used in physics in a thousand years."

I was very skeptical about the practical applications of contemporary mathematical research. The past twenty years have changed my mind.

All the technological objects that you use in your day-to-day life are designed and built using advanced mathematics. Every piece of information that is recorded or transmitted over a distance can only be so thanks to sophisticated mathematical processing. Every time you interact with your smartphone, you interact with interwoven stacks of mathematical abstractions.

For centuries, math has played a prominent role in science and technology. The digitization of our world and our lives has amplified this phenomenon by orders of magnitudes. There's no possible doubt that math is technologically useful, and getting more and more so by the day.

Math already is, in fact, frighteningly useful.

However, when put in a broader historical perspective, the mathematization of the world appears as a recent phenomenon. It dates back to Descartes and just before him Galileo, who famously declared that the universe was a book "written in the language of mathematics."

Prior to the seventeenth century, science wasn't mathematized. Mathematics didn't really have any application. It still consisted of the "childish and pointless" exercises in arithmetic and geometry that

Descartes was so upset about. That hadn't prevented the ancient Greeks from making it a prerequisite for philosophy.

No offense to Galileo, but the notion that mathematics is the language of the universe doesn't make any sense to me. I'm equally skeptical when I hear that math is useful primarily through its applications in science and technology. This way of talking about math makes the whole history of science totally incomprehensible: How did mathematicians find a way to get acquainted with the language of the universe? Did mathematics fall from the sky? Was it sent by God? Why did the ancient Greeks, who hadn't figured out that it was the language of the universe, still insisted on teaching mathematics?

What could have motivated its development over those millennia when it served no practical purpose?

The True Mathematics

Presenting mathematics as an external tool is the surest way to make us hate it. Official math, with its sharp edges, its cold logic, its unbearable air of superiority, is impossible to fall in love with. But as we've seen, there is another way.

I've spoken in this book about how I used my intuition to get ahead in math. At least that's what I believed I was doing at first, when I still thought that what counted was the official math, the stuff in the books.

As I matured, I came to the realization that it worked the other way around. I was using math to develop my intuition.

Math is first and foremost an inner tool. Its main purpose is to enhance human cognition. With the correct exercises of imagination, we have the ability to develop an intuitive and familiar understanding of mathematical notions. We can appropriate them and make them an extension of our bodies.

The true math is the secret math, the one that extends our intuitive understanding of the world that surrounds us.

You already have access to this inner math. You know how to manipulate a circle in your head. You sense the presence of the number 999,999,999 right there in front of you. When you look at the world, you recognize numbers and geometric shapes.

In your head, mathematical concepts behave differently than other concepts. They're much more difficult to learn. But once in place, they provide you with mental images of an incomparable clarity and stability. This is made possible by the unique properties of mathematical truth and logical formalism.

You were a child when you first learned about elephants. Then you learned that there are two different kinds and you could tell them apart by the size of their ears. Now you know that there are three distinct species of elephants. Who knows how many there'll be tomorrow?

With the number 2, you'll never run into such issues. Mathematical truth ties mathematical concepts together to form a mental matrix that is uniquely coherent and stable. It may be difficult to explain what the number 2 really is, but you know that 2 + 2 = 4 and that it's not going to change.

Your mathematical intuition will never become perfect, but logic and mathematical truth enable you to continually refine and recalibrate it.

Even if you think you're terrible at math, the conceptual matrix formed by the math that already lives inside your head is the most solid anchor point of your relationship to the world. Without numbers, without circles and squares, without your perception of points and trajectories in a three-dimensional space, without *x* and *y*, without the concepts of distance, speed, and acceleration, without the idea that a straight line can continue infinitely, without probabilities, without addition and multiplication, without the very notion of truth and logical reasoning, the whole world around you would suddenly become so blurred and unsteady that you'd feel like you'd been lobotomized.

The math that you understand augments reality and adds a magical layer of intelligibility. It makes you hyperlucid.

With time, this math has become so concrete and obvious to you, so "real," that it no longer feels like math. By comparison, the math that you don't yet understand will always seem abstract, absurd, "imaginary."

And yet these concepts that seem so evident and deeply embedded in you weren't always in the picture. It's difficult to believe, but things as simple as whole numbers required people to seek them out, through the power of thought, in the confines of human understanding. They first felt them hatching in the fog of their intuition. Then they struggled to put words to them. They worked to make these words simple and accessible, so that everyone could end up seeing them clearly.

Inside the heads of mathematicians today, there's a thousand times more than all you've been taught.

Mathematics isn't the language of the universe. It's the language that allows us to speak with clarity and precision of all the things that we can't point to with our fingers. It's the language that makes us capable of reasoning and doing science. It's the language that's made us what we are, for better and for worse.

This way of approaching math, as a technique of mental reprogramming and extension of human perception, is fairly recent. It's a vision that had been in the air for some time without anyone taking the time to clarify it and make it accessible to the general public—at least until quite recently.

It is beautifully expressed in these striking lines written by Thurston in 2011: "Mathematics is commonly thought to be the pursuit of universal truths, of patterns that are not anchored to any single fixed context. But on a deeper level the goal of mathematics is to develop enhanced ways for humans to see and think about the world. Mathematics is a transforming journey, and progress in it can be better measured by changes in how we think than by the external truths we discover."

A Work of Fiction

There remains one crucial point, perhaps the most important of the book.

The weird feeling that had troubled me throughout the years had nothing to do with the question of what math is good for.

People who ask that question don't do math. People who do math know quite well it's good for something, if only to give them pleasure, that magical feeling of seeing the world become more and more illuminated the further they progress in math.

The weird feeling had more to do with the actual experience of becoming a mathematician and what it felt like internally. It wasn't like anything I was prepared for. Something really strange was going on.

I'm going to explain what it was, but before I do, I first need to make a few observations on what is quite possibly the most disconcerting aspect of math: the constant reference to things that don't exist "for real" and that you have to try to imagine anyway.

The most simple and fundamental advice you can give to people who want to understand math, which I've repeated throughout this book, is to pretend the things are really there, right in front of you, and that you can reach out and touch them.

People who don't understand math are basically stuck in a state of disbelief. They're refusing to imagine things that don't actually exist, because they don't see the point. It just makes no sense to them.

I admit that it's disconcerting, but the only way to give meaning to mathematics is to imagine that the things it's talking about really exist. Grothendieck is very transparent about it in this passage, which I've previously cited: "All my life I've been unable to read a mathematical text, however trivial or simple it may be, unless I'm able to give this text a 'meaning' in terms of my experience of mathematical things, that is unless the text arouses in me mental images, intuitions that will give it life."

As we've seen, there's nothing grandiose or sophisticated about these mental images. They're always childish, always simplistic, and almost always plainly wrong. When mathematicians think about spheres, they imagine them more or less the same way you do.

Mathematicians are human beings. They can understand mathematical objects only in a perceptual manner, via *false* human interpretations, approximations, translations from mathematical vocabulary into human language.

In fact, this is precisely why math is so beneficial for us: it forces us to enrich our *human* vocabulary and our *human* perception.

On the other hand, mathematicians always keep in mind that their mental pictures are only an approximation of the truth, and they're constantly looking to find out how their pictures are false.

Real spheres exist elsewhere, in a sort of parallel universe. Knowing whether or not this parallel universe really exists is a useless debate, since it's inaccessible anyway. Some mathematicians are convinced that it exists, others are convinced that it doesn't—and still others, like me, couldn't care either way.

The only thing that counts (and this is where it really becomes disconcerting) is that you must imperatively act "as if" this parallel universe existed, because if you don't mathematics is nothing more than a bunch of cryptic symbols on a piece of paper.

This explains the insistence of mathematicians on speaking of *mathematical objects* to designate what most people call *mathematical abstractions.*

In other words, from a purely practical standpoint, math is indistinguishable from fiction.

Learning math is an activity of pure imagination. We bring mathematical objects into our heads through the power of thought and keep them together there through the cohesive effect of a mysterious ingredient, which in a way is the true hero of the fiction: mathematical truth.

Of all the mathematical concepts, truth is at once the simplest and the most difficult to explain. If you want to explain the number 2, you can hold up two oranges. If you want to explain what a triangle is, you can point to a triangle. But what can you hold up or point to in order to explain what truth is for mathematicians?

Believe it or not, the fiction works. Mathematicians develop new ways of approaching reality and new ways of thinking that, throughout history, have demonstrated their effectiveness.

The objects of the fiction, via their concrete and intuitive incarnation, become new concepts that enrich our understanding of the world. It's as if they stepped out of the fiction to become "real," incarnate, like the number 2 becomes real when we see two oranges.

In the process of returning to reality, all mathematical objects lose their perfection, but they conserve the essential characteristics that made them what they were in the fiction. An orange may not be a real sphere, but it's still round.

All mathematical objects, that is, save one. The central character stays stuck in the fiction, as nothing in the real world even remotely resembles mathematical truth.

The moment the dream fades away, mathematical truth instantly vanishes, like a genie going back into its bottle.

An Imaginary Friend

I haven't yet explained what was causing my weird feeling, but it's quite possible that you're starting to experience a weird feeling of your own.

And, to be honest, there is indeed something weird. Not only is there a disconnect between the assumed "rationality" of mathematicians and the strangeness of what they actually do, but the deeper we dig, the stranger it becomes.

I am not aware of any other human activity that involves such a violent back-and-forth between reality and fiction. Seen in this light, the approach as a whole seems utterly insane and doomed from the start. It's a bit like mathematicians were having conversations with an imaginary friend who lets them in on secrets about the world around them. How could this have the slightest chance of success?

This intrinsic weirdness has clouded the understanding of mathematics throughout its history.

There are so-called "imaginary" numbers that are neither more nor less imaginary than so-called "real" numbers, which are neither more nor less real than so-called "rational" numbers.

Each time a new type of number was introduced, it provoked a lot of unease, not only among the public but also among mathematicians themselves, including those who had introduced the new numbers.

In the nineteenth century, there were still serious mathematicians who claimed that negative numbers were nothing but a fairy tale. In the fifteenth and sixteenth centuries even their advocates labeled them *absurd numbers.* Since then, it's as if reality itself had changed, and decided to switch sides. These previously absurd numbers have become concrete and familiar. They've taken over everyday life. To prove to yourself that negative numbers aren't fairy tales, you just have to open a bank account.

As we saw, Cantor was labeled a "scientific charlatan," a "renegade," a "corruptor of youth" for having talked about infinity calmly and precisely. What people really reproached him for was having made tangible what should have stayed evanescent. From a theological perspective, mathematics is unfair competition.

"The essence of mathematics is its freedom," declared Cantor. The freedom of mathematicians is to treat "imaginary" things as "real" things from the moment they are "true." In the end they even see them as being "obvious."

It happens that this approach works remarkably well. Mathematicians obviously aren't going to stop when things are going so well. They continue to amuse themselves with the supernatural or miraculous nature of their constructions. They manipulate "ideals" and "vanishing spectra." The famous "monstrous moonshine conjectures" (named after the "Monster," an object that lives in dimension 196,883) were proved using a "no-ghost theorem." In algebra, there's a construction called the *Eilenberg swindle.*

If the process of understanding math is already quite bizarre, the discovery process is even more so. The experience is so singular and disconcerting that most accounts look like they were written by mystics.

One of the most baffling aspects is the abrupt manner in which ideas come to you, without effort and almost always inconveniently. They emerge, as Grothendieck puts it, "as if summoned from the void."

In an influential research article by Bob Thomason and Tom Trobaugh, we're told that the second author contributed only after he was dead, by means of appearing in a dream of the first author. Not only did he suggest the right approach, he stopped the first author from readily dismissing it as hopeless: "Tom's simulacrum had been so insistent, I knew he wouldn't let me sleep undisturbed until I had worked out the argument."

One of my close friends, an excellent mathematician whose name I won't disclose, recently told me that he had the distinct impression (which he never dared share with others) that the greatest ideas in his career had been directly suggested by God (even though he's an avowed atheist).

For my part, I've never felt anything along those lines. I've simply had the impression of being able to levitate and pass through walls.

Well-Guarded Secrets

That was the weird thing. The more I advanced, the further I dove into the heart of mathematics, the more I learned to master the techniques that facilitate deep understanding and creativity, the more it began to resemble witchcraft and black magic.

Descartes thought that mathematicians guarded their secrets for fear of losing their prestige. If people knew that there was a method and it was that simple, he reckoned, they would stop looking at mathematicians like they were demigods, and come to the realization that they're just normal people.

The real explanation is undoubtedly more trivial: mathematicians are simply afraid of being called insane.

If I hadn't become one of them myself, I might have continued to believe that they were demigods capable of speaking the language of the universe. But I know it's not true. I know where I come from. I saw what made me get better. Each key step was always the more or less fortuitous discovery of a new technique to overcome my inhibitions or a new way of making my imagination work.

In practice, mathematics doesn't have much to do with the hard sciences. It's rather more related to psychology, of which it's a kind of esoteric and applied sub-branch.

It's undeniable that mathematical creation feels magical and supernatural. But behind all that there's necessarily a human reality that is neither supernatural nor magical.

What really troubled me and made me want to continue to explore these subjects until I felt able to tell the story in a simple manner was an immense feeling of waste.

No other human project has the prestige and intellectual authority as that of mathematics. If mathematicians are incapable of explaining their approach without giving the impression that they're some kind of shamans, that doesn't mean they're actual shamans.

It simply means that they're not using the right words and that their explanation is incomplete.

The Right Way of Waving My Hands

Why is it so hard to teach math? Why has this been so for centuries and centuries? What do we fail to share and communicate? What exactly are we missing?

I studied math because I couldn't understand how it was possible to understand it. I expected someone would explain to me *why* it was possible and *how* to do it. The explanation never came. The subject was never even raised.

That didn't stop me from learning on my own. Like so many others, I experienced the frustration of staying silent about the aspect of mathematics that, for me, was of the most value.

Each time I found myself in a situation of teaching or explaining my work, I tried to bring together two levels of discourse: a formal level made of rigorous definitions and precise statements, and an intuitive level, with the right metaphors, the right drawings, the right inflection of my voice, the right way of waving my hands.

These two levels complement one another. A formal lecture without motivation and without sharing your intuition is meaningless. But a purely intuitive discourse without any formalization is equally meaningless—this is why attempts at popularizing math so often miss the mark. Once you get rid of official math, intuition loses its moorings.

There is an inherent limit to how much math can be taught without formal definitions and formal statements. At some point, without formalism, it's no longer mathematics, it's just people waving their hands.

Shortly before quitting academia, I had the opportunity to give the most interesting course of my career. It was a semester-long introductory math course for literature and philosophy students at the École normale supérieure, one of the most prestigious universities in France.

It was an opportunity to experiment, and confront myself with this fundamental question: can you teach the art of seeing math in your head?

I plunged back into what are traditionally called the *foundations* of mathematics: logic and set theory. That's when I realized that I had been going about it the wrong way. Logic and set theory are not *foundations* of mathematics, they are *branches* of mathematics. Their focus is the mathematical formalization of the notion of proof—a perfectly legitimate field of study, but one that won't bring much clarity to what math really is, let alone how to teach it.

Certain ideas and examples in this book go back directly to my class notes from this time. Back then, however, I was missing a crucial ingredient.

In my class, I constantly felt that something wasn't right, as if I hadn't managed to situate the conversation in the right place. I loved the math that was alive inside of me, but I was unable to explain it in words that others could relate to.

It was in this context that I called an end to my scientific career. Decisions of this kind are never easy to make. Trying to pinpoint a single factor that explains it all would be naïve. Among the multitude of factors, there was, however, this particular frustration: I wasn't able to teach math in a meaningful way. I could teach what math was supposed to be, but I couldn't teach what it really was for me. Somehow, it felt as if this sincere level of teaching wasn't permitted, as if an ancient taboo was preventing it from happening.

With hindsight, it's now clear that my introductory course was simply missing a discussion of the human experience of understanding mathematics.

The Impossible Story

Euclid's *Elements* is the most influential mathematics treatise in history. It dates back twenty-three hundred years and, throughout the centuries, it has come to serve as a blueprint for mathematical reasoning itself.

Since this time, mathematics has been presented as the science of logical deduction. The other part of the story, that which concerns the unseen actions that we perform in our heads, has been obscured. This is of course no conspiracy. A more plausible explanation is that, for twenty-three hundred years, this other part of the story was simply impossible to tell.

Telling the story would have required explaining what was going on in our heads, and we had no satisfying way of representing the inner workings of our own intelligence.

The only model we could think of was that of mechanical deductive reasoning, in the spirit of Euclid's *Elements.* Viewing intelligence as the ability to perform *calculations* is as old as math itself. "Calculation" comes from the Latin *calculus,* which means "small pebble," referring to the stones used on an abacus for counting. Across the centuries, the brain was compared to an abacus, then a geared computing machine, then a silicon chip. The metaphor conflated mechanical deductive reasoning with mathematics and rationality—and with intelligence itself.

As we've seen, this metaphor is deeply flawed. This profound misinterpretation of our brain processes has made us unable to relate mathematics to the common human experience.

Ironically, we've always known that our intelligence couldn't be reduced to calculations. We've known deep inside that something else was going on. Yet we had no way to evoke it without having recourse to the supernatural. We were stuck with *spirits* and *intuitions, third eyes* and *sixth senses.* We imagined magical entities acting outside of our control, and with which only a small elite, endowed with a special gift, had the privilege of being able to directly communicate. These models had remained practically unchanged since prehistory.

Our language itself was a mystery. Who invented words? What is a concept? What is meaning? What is truth? How are we able to even make sense of phrases? For millennia, these weren't even science questions—they belonged to the fields of metaphysics and theology.

To understand math is to reprogram your intuition. It is, above all, a matter of neuroplasticity. The secret techniques of mathematicians are neither more nor less paranormal than those that allowed Ben Underwood to see the world by clicking his tongue.

As long as we treated our mental activity as something magical, mathematics was fundamentally impossible to explain. But as we're entering the age of artificial intelligence, we may finally have an opening.

It was my encounter with deep-learning algorithms that enabled me to write this book. For the first time in my life, I had access to a brain metaphor that allowed me to make sense of my own journey. I realized that my testimonial had value, as I could describe my subjective experience in terms that were sufficiently clear, sufficiently "rational," to escape the safe confines of private conversations.

This is how I ceased to view my story as an impossible story.

When I look at them with the deep-learning metaphor in mind, the strange phenomena that have troubled me for so many years suddenly cease being strange to me. Yes, ideas come unexpectedly, "as if summoned from the void," but that's normal. Yes, plasticity is a slow and silent mechanism that occurs without any real effort on our part, provided we're exposed to the right mental images. Yes, we learn precisely when we force ourselves to imagine things *that we don't yet understand,* which unfortunately is the same exact thing that most people run away from. Yes, paying attention to the small details that trouble us is of the utmost importance, and the fastest way to learn is to follow the path of maximum perplexity. Cartesian doubt, within this framework, can be interpreted as an "adversarial" hack to accelerate our learning.

A Mathematical Awakening

For millennia, we presented mathematics in a way that made it unintelligible to most people. We finally have the opportunity to talk about it differently.

Learning math should be like learning any other motor skill, like learning to swim or ride a bike, and it should be accessible to everyone. Our false beliefs about the nature of our language and the functioning of our thought are obstacles to this simple and direct learning. They instill fears and inhibitions that block the unseen actions without which no mathematical learning can take place.

How do you teach math to someone who believes that their intuition and perception of reality are given and impossible to reprogram? It's exactly like teaching swimming to someone who is convinced their body is as dense as rock and will sink. A prelude to any successful teaching is getting rid of such beliefs.

This is why I conceived of this book as a book of awakening and emancipation. I believe that mathematics has taught me important lessons about our bodies, how they function, and what we can accomplish with them, and I wanted to share those lessons.

I've tried to speak as candidly as possible about what goes on inside our heads: the subjective reality, the emotional journey, the physical and sensorial experience, the practical aspects of what we do, how we do it, how it works, and what it feels like. None of that has ever been part of teaching, because it wasn't supposed to be part of math, and also because of our deeply ingrained suspicion toward anything subjective.

I'm well aware of the shortcomings of subjectivity. "I know how likely we are to be wrong on our own account," wrote Descartes, and the situation hasn't improved since his time.

But do we really have a choice? When mathematics was just about proving theorems, it was entirely natural to treat the subjective experience as a second-rate citizen, to be covered only informally and anecdotally, if time allowed. But as soon as we realize that mathematics is ultimately about human understanding, this ceases to be an option. Understanding is, in essence, a subjective experience.

My personal story would be of little value if it was simply that of "one with a gift." While I cannot prove that I have no special talent, I know for a fact that I was myself extremely intimidated by the notion that creative mathematicians had to be biologically different, and would have given up if I hadn't found a way to get rid of it.

From the outside, it indeed seems that mathematical creativity requires some sort of superhuman intelligence. To stop thinking in terms of "gifts" and "talents," one has to find an alternate explanation.

My way of looking at things, which has served me well throughout my career, was to imagine that creative mathematicians were hackers who had found ways to unlock "hidden modes" of our cognition. Most of the time, they'd done so unwittingly, and were entirely incapable of explaining how.

It's also the hypothesis that's guided me throughout the writing of this book. I concentrated on the few things that I knew firsthand, and thus on a small part of the subject.

I was fortunate enough to be able to rely on the writings of Descartes, Grothendieck, and Thurston. Their stories closely resemble one another, as if it were the same story told from three different points of view. These stories are compatible with what I've lived through myself, which allowed me to inscribe my personal journey within a tradition that's older, stronger, and more richly documented.

Even they didn't have all the keys. Descartes was unable to account for what happened to him without making the *dualist* assumption that his mind was of a divine and immaterial essence, detached from his body. Similarly, Grothendieck was convinced that God whispered in his ear and dreamed inside his head. Thurston is the most pragmatic of the three, the most modern, and without doubt the most lucid.

Their candor and sense of detail are of utmost value. They tried their best to share what they'd lived through and what, in their view, accounted for their success.

From the first page to the last, they speak almost only of imagination. Each describes the use of imagination in new modalities, discovered by accident and breaking with what they'd been taught.

Grothendieck attributes the singularity of his work to his transgression of a taboo: "It would seem that among all the natural sciences, it is only in mathematics that what I call 'the dream' or 'the daydream' is struck with an apparently absolute interdiction, more than two millennia old."

Thurston puts it in a less grandiose but equally impactful fashion: "I have decided that daydreaming is not a bug but a feature."

Descartes, himself an avid daydreamer, left us with a compelling account of his powerful techniques. But, by a cruel irony, he's also responsible for the flawed theory that is making it so hard for us to take him seriously.

Imagination is the key to this entire story. Mathematicians have found a unique way to use theirs and it has made them incredibly successful. Mathematical imagination is all the more visionary and limitless in that it is guided by *mathematical truth,* the secret ingredient that makes it possible to figure out *the right things to imagine,* the ones that will eventually solidify and expand our intuitive understanding of the world around us. This is where the true foundations of mathematics are to be found, and not in formal logic or theology.

But under a narrow-minded version of rationalism, none of this can make any sense.

We've been taught that our thoughts have no impact on reality. For an educated person, it takes a lot of guts to affirm otherwise. We read here and there that one can change the world by the power of thought, but we're trained to dismiss it as mysticism or self-help bluster.

This is dualism at its finest. Our thoughts are presumed to be ethereal. You can imagine whatever you wish, however you wish, without that having any impact whatsoever on the physical world.

Yet using our imagination isn't navigating an ethereal layer of the cosmos. Nor is it a parasitical activity that we should seek to suppress. It is, instead, a genuine physical activity that is central to human cognition.

What we see and do *in our heads* contributes to neural learning every bit as much as what we see and do *for real.* If we feel the urge to perform unseen actions in our heads, if we dream and if we daydream, it's because this allows us to *fabricate understanding.* What we imagine modifies the actual wiring of our brain and literally *changes the way we see the world.*

There are a thousand and one ways to imagine. We haven't yet learned to recognize them all, and still less to name them. *Think, meditate, reflect, visualize, analyze, fantasize, reason, dream:* we use these words haphazardly, without really knowing what they mean, and without realizing how much they have in common.

It's through this vagueness that all the misunderstandings slip in. No one cared to even tell us that there are right and wrong ways of using our imagination. Some make us stupid. Some make us crazy. And some have the power of making us incredibly smart.

Now that we're teaching machines the secrets of intelligence, it's about time we start teaching humans.

Epilogue

Early in 1913, G. H. Hardy, an eminent mathematician at the University of Cambridge, received a strange letter from Madras, India.

The writer was a man named Srinivasa Ramanujan. He said he was a twenty-three-year-old clerk living in poverty, without any higher education, who spent his free time studying mathematics on his own. He accompanied his letter with a selection of theorems he claimed to have obtained and that the local mathematicians had deemed "surprising." He was curious to hear Hardy's opinion.

Hardy gave the theorems a quick glance. He thought at first it was some kind of a hoax. However, the more he looked at the manuscript, the more perplexed he became. Not only did the results seem credible, their depth and originality were extraordinary, and Hardy felt completely blown away.

The theorems were given without proofs. Hardy himself was incapable of proving them. He thought, however, that "they must be true because, if they were not true, no one would have had the imagination to invent them."

Hardy therefore concluded that Ramanujan was a mathematician of the highest order, who would take his place among the greatest in history.

Formalism and Intuition

The story of the encounter and friendship between Hardy and Ramanujan is so improbable that you'd think it was taken from fiction.

It can be read as a social fable. At the height of British colonial domination, two worlds collide. Hardy is a pure product of Western intellectual arrogance, a member of the most elite circles, comfortably ensconced in his ivory tower. Ramanujan is a self-taught amateur mathematician, the son of a sari vendor.

Hardy invites Ramanujan to Cambridge, where he lives for five years, from 1914 until 1919, when, gravely ill, he decides to return to India, where he dies the following year at the age of thirty-two.

At the end of his career, when Hardy was asked about his greatest contribution to mathematics, he replied without hesitation "the discovery of Ramanujan."

Hardy had reason to be proud. He had immediately recognized the extraordinary genius of Ramanujan. He'd had the courage and integrity to act on this even though it meant going against established norms. Ramanujan was the first Indian to be elected as a fellow of Trinity College and one of the youngest fellows of the Royal Society.

On another level, the story can also be read as a mathematical fable. It reprises the principal themes we've addressed in this book and constitutes the perfect epilogue.

From the beginning of this book, we've talked about how mathematics feeds on the tension between two contradictory forces: the inhuman coldness of logical formalism and the phenomenal power of intuition. All mathematical work, whether the resolution of a primary school exercise or research that extends the boundaries of human knowledge, requires a constant dialogue between formalism and intuition.

Not everyone approaches this dialogue in the same manner. Some mathematicians are spontaneously more "formalist," while others are more profoundly "intuitive." Yet they all know that in order to progress they need to reach out to both sides.

The duo act formed by Hardy and Ramanujan is all the more fascinating in that they are perfect incarnations, almost to the point of caricature, of these two polarities.

Hardy was one of the most famous mathematicians of his time and one of the principal figures in the formalist revolution that, at the beginning of the twentieth century, allowed for the unification of mathematics and the formalization of the notion of proof.

Hardy was a friend of Bertrand Russell, the coauthor (along with Alfred North Whitehead) of the most inhuman book in the history of thought: *Principia Mathematica.* In an ultra-formalist style verging on the delirious, this treatise (whose title alludes to Newton's great work) provides axiomatic foundations to set theory, solidifying Cantor's initial vision and demonstrating along the way that the concept of numbers can be reconstructed from the concept of sets.

This monumental work changed the face of mathematics. Conceived for the ages, it was unfortunately disfigured by a nasty birth defect: it was indecipherable to any person of normal understanding. If you're looking for the proof that $1 + 1 = 2$, you'll find it on page 379.

Upon the publication of *Principia Mathematica,* Hardy wrote a review for the general public that appeared in the *Times Literary Supplement.* With characteristic British humor, he stated, "Non-mathematical readers may very naturally be frightened by an exaggerated notion of the technical difficulty of the book."

As for Ramanujan, he was the most phenomenally intuitive mathematician in history. It is difficult to speak of him without recourse to superlatives. Our vocabulary simply isn't adequate. Even the word *genius* seems too feeble.

The way he worked defies understanding. He simply wrote down bizarre formulas on pieces of paper headed by the word *theorem* without giving the least explanation of his thought processes.

When Hardy insisted upon the necessity of coming up with rigorous proofs, Ramanujan responded that he didn't see the need. He knew that the formulas were correct, because his family's goddess Namagiri Thayar had revealed them to him in a dream.

I would have loved to have been a fly on the wall to see the expression on the face of Hardy, a confirmed atheist and fervent rationalist, when Ramanujan dared say such things.

In the course of his short career, Ramanujan produced more than thirty-nine hundred "results." What status should they be given? Normally, a theorem without a proof is not a theorem but simply a conjecture. In any case, that's the official version.

A century after his death, his confirmed record is prodigious. Almost all of his formulas have been shown to be correct. The search for proofs has inspired the development of entire fields of mathematics and required the invention of sophisticated new conceptual

***54·42.** $\vdash :: \alpha \in 2 \,.\, \supset :. \beta \subset \alpha \,.\, \exists ! \beta \,.\, \beta \neq \alpha \,.\, \equiv \,.\, \beta \in \iota``\alpha$

Dem.

$\vdash \,.\, *54\cdot4 \,.\quad \supset \vdash :: \alpha = \iota`x \cup \iota`y \,.\, \supset :.$

$\qquad \beta \subset \alpha \,.\, \exists ! \beta \,.\, \equiv : \beta = \Lambda \,.\, \vee \,.\, \beta = \iota`x \,.\, \vee \,.\, \beta = \iota`y \,.\, \vee \,.\, \beta = \alpha : \exists ! \beta :$

$[*24\cdot53\cdot56 . *51\cdot161] \qquad \equiv : \beta = \iota`x \,.\, \vee \,.\, \beta = \iota`y \,.\, \vee \,.\, \beta = \alpha \qquad (1)$

$\vdash \,.\, *54\cdot25 \,.\, \text{Transp} \,.\, *52\cdot22 \,.\, \supset \vdash : x \neq y \,.\, \supset \,.\, \iota`x \cup \iota`y \neq \iota`x \,.\, \iota`x \cup \iota`y \neq \iota`y :$

$[*13\cdot12] \quad \supset \vdash : \alpha = \iota`x \cup \iota`y \,.\, x \neq y \,.\, \supset \,.\, \alpha \neq \iota`x \,.\, \alpha \neq \iota`y \qquad (2)$

$\vdash \,.\, (1) \,.\, (2) \,.\, \supset \vdash :: \alpha = \iota`x \cup \iota`y \,.\, x \neq y \,.\, \supset :.$

$\qquad \beta \subset \alpha \,.\, \exists ! \beta \,.\, \beta \neq \alpha \,.\, \equiv : \beta = \iota`x \,.\, \vee \,.\, \beta = \iota`y :$

$[*51\cdot235] \qquad \equiv : (\exists z) \,.\, z \in \alpha \,.\, \beta = \iota`z :$

$[*37\cdot6] \qquad \equiv : \beta \in \iota``\alpha \qquad (3)$

$\vdash \,.\, (3) \,.\, *11\cdot11\cdot35 \,.\, *54\cdot101 \,.\, \supset \vdash \,.\, \text{Prop}$

***54·43.** $\vdash :. \alpha, \beta \in 1 \,.\, \supset : \alpha \cap \beta = \Lambda \,.\, \equiv \,.\, \alpha \cup \beta \in 2$

Dem.

$\vdash \,.\, *54\cdot26 \,.\, \supset \vdash :. \alpha = \iota`x \,.\, \beta = \iota`y \,.\, \supset : \alpha \cup \beta \in 2 \,.\, \equiv \,.\, x \neq y \,.$

$[*51\cdot231] \qquad \equiv \,.\, \iota`x \cap \iota`y = \Lambda \,.$

$[*13\cdot12] \qquad \equiv \,.\, \alpha \cap \beta = \Lambda \qquad (1)$

$\vdash \,.\, (1) \,.\, *11\cdot11\cdot35 \,.\, \supset$

$\qquad \vdash :. (\exists x, y) \,.\, \alpha = \iota`x \,.\, \beta = \iota`y \,.\, \supset : \alpha \cup \beta \in 2 \,.\, \equiv \,.\, \alpha \cap \beta = \Lambda \qquad (2)$

$\vdash \,.\, (2) \,.\, *11\cdot54 \,.\, *52\cdot1 \,.\, \supset \vdash \,.\, \text{Prop}$

From this proposition it will follow, when arithmetical addition has been defined, that $1 + 1 = 2$.

***54·44.** $\vdash :. z, w \in \iota`x \cup \iota`y \,.\, \supset_{z,w} \,.\, \phi(z, w) : \equiv \,.\, \phi(x, x) \,.\, \phi(x, y) \,.\, \phi(y, x) \,.\, \phi(y, y)$

Dem.

$\vdash \,.\, *51\cdot234 \,.\, *11\cdot62 \,.\, \supset \vdash :. z, w \in \iota`x \cup \iota`y \,.\, \supset_{z,w} \,.\, \phi(z, w) : \equiv :$

$\qquad z \in \iota`x \cup \iota`y \,.\, \supset_z \,.\, \phi(z, x) \,.\, \phi(z, y) :$

$[*51\cdot234 . *10\cdot29] \equiv : \phi(x, x) \,.\, \phi(x, y) \,.\, \phi(y, x) \,.\, \phi(y, y) :. \supset \vdash \,.\, \text{Prop}$

***54·441.** $\vdash :: z, w \in \iota`x \cup \iota`y \,.\, z \neq w \,.\, \supset_{z,w} \,.\, \phi(z, w) : \equiv :. x = y : \vee : \phi(x, y) \,.\, \phi(y, x)$

Dem.

$\vdash \,.\, *5\cdot6 \,.\, \supset \vdash :: z, w \in \iota`x \cup \iota`y \,.\, z \neq w \,.\, \supset_{z,w} \,.\, \phi(z, w) : \equiv :.$

$\qquad z, w \in \iota`x \cup \iota`y \,.\, \supset_{z,w} : z = w \,.\, \vee \,.\, \phi(z, w) :.$

$[*54\cdot44] \qquad \equiv : x = x \,.\, \vee \,.\, \phi(x, x) : x = y \,.\, \vee \,.\, \phi(x, y) :$

$\qquad y = x \,.\, \vee \,.\, \phi(y, x) : y = y \,.\, \vee \,.\, \phi(y, y) :$

$[*13\cdot15 \qquad \equiv : x = y \,.\, \vee \,.\, \phi(x, y) : y = x \,.\, \vee \,.\, \phi(y, x) :$

$[*13\cdot16 . *4\cdot41] \equiv : x = y \,.\, \vee \,.\, \phi(x, y) \,.\, \phi(y, x)$

This proposition is used in *163·42, in the theory of relations of mutually exclusive relations.

tools. This work involved mathematicians of the first order over decades upon decades. We're only just now beginning to have the end in sight.

How did Ramanujan discover his formulas? Wasn't his way of seeing them the beginning of a proof, if not a complete but nonverbal proof? Did he really have no way of saying more about it without invoking his goddess?

Under Hardy's influence, Ramanujan was able to learn the rudiments of "academic" mathematics. He finished his thesis and wrote a couple of articles that contained actual proofs. However, he never succeeded in explaining his work method. If he had lived longer, perhaps he would have been able to find a way to better explain the images, colors, or structures, the tastes or the textures that formed inside his head, and how he learned to invoke them.

If you really want to believe in magic or the existence of supermen with supernatural powers, you might find some inspiration in the story of Ramanujan.

As for me, I'm siding with Misha Gromov, one of the greatest living mathematicians (he received the Abel Prize in 2009). For Gromov, it would be a mistake to attribute Ramanujan's genius to some cosmic anomaly, a singularity cut off from common human experience: "This miracle of Ramanujan forcefully points toward the same universal principles that make possible mastering native languages by billions of children."

I suspect that Gromov's affirmation is born of his personal experience, of the intimate understanding he has of the mechanisms of his own creativity, which is itself miraculous enough.

As we near the end of this book, I hope that a remark like Gromov's no longer comes as a surprise to you, and that it even seems quite natural.

Someone Exactly Like You or Me

In Hardy's review of *Principia Mathematica,* lurking behind his British sense of humor, you can discern a less sympathetic aspect of his personality: his morbid elitism.

The review was written for a general audience, the readers of the *Times,* on a subject they had legitimate reasons to find intimidating: a 666-page book with a Latin title that aspired to serve as a new foundation to logic, mathematics, and human thought as a whole.

While admitting that "the general tone of the book is mathematical," Hardy prefers to insist on its philosophical implications and historical character, employing a lighthearted tone that makes you think he enjoyed delving into it.

He correctly notes that the book contains "crazy looking symbolism" and that "it would be silly to pretend that the book is not really difficult," but that doesn't prevent him from asserting, "It has many claims to be widely read." Hardy goes so far as to say that "some of the jokes are very good."

At no point does he reveal the key to the enigma, the crucial advice my friend Raphael gave me that I shared in chapter 6. To anyone confronted with *Principia Mathematica,* this advice becomes a basic mental health tip: "Never ever read math books."

For Hardy, mathematics was a gentlemen's club that only the select few could enter. In his famous autobiography, *A Mathematician's Apology,* a book that was once considered a classic but, to modern readers, seems plainly obnoxious, Hardy goes so far as to proclaim this malediction: "There is no scorn more profound, or on the whole more justifiable, than that of the men who make for the men who explain. Exposition, criticism, appreciation, is work for second-rate minds."

The brutal elitism of the mathematical community is unfortunately a topic on its own. It's a tradition that goes back centuries.

In the academic world, mathematicians build their careers and legitimacy upon the new theorems they prove, and upon nothing else. All the rest doesn't really count, apart from the rare conjectures that become famous in themselves and confer a special prestige.

This system has its merits. It reduces arbitrariness and helps mathematicians guard against complacency and nepotism. When a discipline deals with eternal truths, it offers a neat way to evaluate careers.

The approach also has its blind spots. Hardy's curse remains very potent and no one is immune to it. In the research community, an exaggerated interest in mathematical education is commonly perceived as a sign of weakness. Fortunately, mathematicians have started to change their opinions in this issue. They have learned to *slightly less* despise teaching and popularization. But there's still a long way to go.

The secret math, the one that deals with human understanding, will never possess the rigor and objectivity of official math. Because of this, it will never be considered as a "serious" topic.

This "nonserious" topic, however, is arguably much more important than most properly mathematical questions.

It concerns anyone who at one moment or another has been faced with learning math—that is to say, absolutely everyone. Mathematicians themselves are passionate about it, and it regularly comes up in their private conversations. It raises fundamental questions about human intelligence, language, and how our brains work.

It would be a terrible mistake to confine this subject to the backstage of science, to late-night conversations and the autobiographies of retired mathematicians. It would equally be a shame to exile it from the field of mathematics and make it the exclusive domain of neurology.

Failing to place human understanding at the center of mathematics is failing to acknowledge the very nature of mathematics.

Undoubtedly we did not have, until quite recently, the tools and the framework to approach the subject in a constructive manner. We were collectively trapped in fatalism and passivity: "Some people in this world are incredibly brilliant at math, but it's no use trying to figure out why, it's simply a miracle, a gift from heaven. And too bad for those who can't understand."

This "nonserious" but burning topic is the subject of this book. I have tried to approach it in my own way, starting from a simple premise: talking about math as I have experienced it, in the simplest way possible, examining what it *really* consists of, the things you do inside your head, and how to approach it *concretely.*

If Hardy had known the right questions to ask Ramanujan, who knows what we might have learned? Fortunately we have the writings of Descartes, Grothendieck, Thurston, and of course Einstein.

It's hard to overestimate the value of these writings. Their most troubling message, the most powerful and subversive, is this: we construct our intelligence on our own, with ordinary human means, with our imagination, curiosity, and sincerity.

Grothendieck wrote: "The man who first discovered and mastered fire was someone exactly like you or me. Not at all what you'd call a 'hero,' or 'demi-god,' or whatever."

Descartes and Einstein said essentially the same thing, albeit in different words. We sliced their brains into sections and put their skulls into museums—but we refused to listen.

What to Do about All of This?

I wrote this as a kind of handbook, something that I would have loved to have on my nightstand during my studies to guide me, to encourage me, to help overcome my inhibitions. I believe it would have helped me immensely. I hope it will help you.

My ambition isn't to make math easier. It never will be, not for anyone. It's not math's job to be easy. I simply want to make it more accessible, to allow those who want to explore it be able to do so, according to their desire and ambition.

There will always be some people who are better at math than others: the visionaries, the passionate, the adventurous. But pretending that being good at math requires a special gift is a lie. Math belongs to all of us. There's no reason to accept being petrified and giving up, neither for ourselves nor for others.

One of the great lessons math has taught me is that it's only by confronting head-on our impression of not understanding something that we have a chance to finally understand it. It seems that scrutinizing our own perplexity is the best way to mobilize our natural faculties for learning.

This is precisely why math is difficult: it requires looking straight at what is beyond our comprehension. We must become genuinely interested in it. We must force ourselves to imagine it and put words to all our impressions, without being distracted by our constant feeling of inferiority. And we must do that precisely when our instinct tells us to run away as quickly as we can.

For Descartes, math is the only place where one can fully experience what it means to *understand* something.

This is also what I personally get from math. Math has taught me to pay attention to that special taste in my mouth, this impression that something's not quite right and doesn't work like it should. It has taught me to recognize a novel idea when it still looks like nothing, to nurture it and pay close attention, so that it has a chance to grow. It has taught me to listen to my emotions.

I know now that my candor and my sensitivity are my most powerful intellectual weapons. The mathematical approach is one of integrity and being in tune with oneself.

Using this approach, I formed habits that I've kept to this day. I stopped believing that there were things that were counterintuitive by nature. The things we're told are "counterintuitive" or "paradoxical" are either false or poorly explained.

No one is forcing us to live in a world that is indecipherable and incoherent. By adopting the right habits and developing our confidence in our ability to *imagine* and to *formalize,* we can continually grow our mental clarity.

If we teach math to children, it's not so much in order to teach them about numbers and shapes as to give them the chance to approach the world in this manner.

Understanding things is one of the great pleasures in life. This pleasure is sometimes spoiled by a feeling of regret for lost time: *How is it I was so stupid not to have understood this earlier?*

I've known this feeling for so long that I no longer pay any attention to it. As the saying goes, the best time to plant a tree was twenty years ago; the second best time is now.

If you think you're terrible at math, and this book makes you want to try again, keep this in mind: there are nice stories about climbing Mount Everest, some even read like novels, but nothing beats training. For beginners, the first few yards of the climb can be the hardest.

My advice is to have no shame in starting at the beginning, with the most classical and elementary proofs. Since it won't be easy for you to know if you really understand them, try explaining them to someone else, perhaps a child.

When we try explaining things to others, we often realize that our own ideas aren't as clear as we thought. It's a painful and humiliating experience, but one that you can get over, and it's precisely by getting over it that you get ahead in math.

My only way of understanding things is to explain them to myself, in the simplest terms possible, as if I were a child. It's the same principle that I used to write this book.

In "On Proof and Progress in Mathematics," Thurston gave this beautiful definition: "Mathematicians are those humans who advance human understanding of mathematics."

That's exactly what I've tried to do.

Notes and Further Reading

Chapter 1. Three Secrets

"I have no special talent. I am only passionately curious." The original sentence, "Ich habe keine besondere Begabung, sondern bin nur leidenschaftlich neugierig," is taken from a letter from Einstein to his biographer Carl Seelig, dated March 11, 1952.

"Do not worry about your difficulties in Mathematics. I can assure you mine are still greater" is taken from a letter from Einstein to Barbara Wilson, a high school student, dated January 7, 1943.

"I believe in intuition and inspiration" is from an interview with Einstein by George Sylvester Viereck published in the *Saturday Evening Post,* October 26, 1929.

Many of Einstein's quotations that you come across are incorrect or altered. Those that I cite in this work have been sourced and verified thanks to *The Ultimate Quotable Einstein,* a collection of verified citations by Alice Calaprice (Princeton, NJ: Princeton University Press, 2011).

Chapter 2. The Right Side of the Spoon

Math is the subject:

—the most difficult for 37 percent of 1,028 American adolescents surveyed by Gallup in a 2004 poll. See Lydia Saad, "Math Problematic for U. S. Teens," *Gallup,* May 17, 2005, or online at https://news.gallup.com/poll/16360/math-problematic-us-teens.aspx.

—the most liked for 23 percent of American adolescents, far more than English (13 percent), according to a study by Gallup in 2004 of 785 Americans aged thirteen to seventeen. See Heather Mason Kiefer, "Math = Teens Favorite School Subject," *Gallup,* June 15, 2004, or online at https://news.gallup.com/poll/12007/Math-Teens-Favorite-School-Subject.aspx.

—the most hated according to countless surveys, of whatever population.

Chapter 3. The Power of Thought

In this chapter, the ability to "see" a circle in one's head is presented as a universal human capability, which isn't entirely correct. In 2015, a research team led by Adam Zeman at the University of Exeter described a rare condition, *aphantasia,* characterized by the inability to create mental imagery. A 2022 study estimates the prevalence of complete aphantasia at around 1 percent. See https://en.wikipedia.org/wiki/Aphantasia for details and references.

When writing this book, I was faced with the challenge of describing what's going on inside our heads. It was natural for me to put an emphasis on the visual experience, as it is easy to convey and most people find it striking. My apologies to readers with aphantasia, who may occasionally feel left out.

One reader with aphantasia reached out after the original edition was published, and I had the opportunity to discuss this chapter with him. While he can't "see" the line sweeping across the circle, he does find "obvious" that a line cannot intersect a circle in more than two points (without being able to provide a reason).

This lone testimonial is by no way a robust scientific study, but it does help illustrate a key point that is reasserted in chapter 16: mathematical intuition comes in many shapes and forms. It doesn't have to be visual.

Chapter 4. Real Magic

The number 999,999,999 would have been easy to write in the Babylonian sexagesimal system invented four thousand years ago, well before the Roman era.

Even though it is difficult to write in Roman numerals, the number itself is easily calculated using an abacus, which the Romans used, and which is implicitly decimal. The problem comes in writing it down outside of the abacus.

After the classical era, Roman numerals were extended to express 1 million, 1 billion, and so on. But writing the number 999,999,999 continues to pose a problem despite these extensions, since you have to use many symbols: just writing "9 million" means you have to use the symbol for million nine times.

Chapter 5. Unseen Actions

The dolphin in the photo is Wave, one of Billie's friends. This photo is taken from the scientific article by M. Bossley, A. Steiner, P. Brakes, et al., "Tail Walking in a Bottlenose Dolphin Community: The Rise and Fall of an Arbitrary Cultural 'Fad,'" *Biology Letters* 14, no. 9 (September 2018), http://dx.doi.org/10.1098/rsbl.2018.0314. This short, accessible article contains a number of interesting details.

"It was not that I was trying to win, but I was trying to not lose." This quote from Fosbury is taken from a 2014 video interview, available on YouTube at https://www.youtube.com/watch?v=gGqQXDkpgss.

"I think quite a few kids will begin trying it my way now. I don't guarantee results, and I don't recommend my style to anyone." Cited in Joseph Durso, "Fosbury Flop Is a Gold Medal Smash," *New York Times,* October 22, 1968.

Chapter 6. Refusing to Read

Thurston's article, "On Proof and Progress in Mathematics," *Bulletin of the American Mathematical Society,* no. 30 (1994): 161–77, is available online at https://arxiv.org/pdf/math/9404236.pdf.

Chapter 7. The Child's Pose

The "ridiculous piece" letter appears in Alexander Grothendieck and Jean-Pierre Serre, *Correspondance Grothendieck-Serre,* ed. Pierre Colmez et Jean-Pierre Serre (Paris, Société mathématique de France, 2001). A bilingual edition was jointly published by the American Mathematical Society and the Société mathématique de France in 2004.

Grothendieck's exact words are "emmerdante rédaction," which is powerfully expressive in a way that is hard to render in polite English ("annoying write-up" is way too mild).

The quotations from Serre are taken from a fascinating conversation with Alain Connes (himself a first-rate mathematician and recipient of the Fields Medal in 1982) at the Fondation Hugot du Collège de France on November 27, 2018. This exceptional insight into the personalities of Grothendieck and Serre is available online at https:// www.youtube.com/watch?v =pOv-ygSynRI.

Unless otherwise noted, Grothendieck's quotations are taken from *Récoltes et semailles: Réflexions et témoignage sur un passé de mathématicien* (Harvests and Sowings: Reflections and Testimonials on a Mathematician's Past), 2 vols. (Paris: Gallimard, 2022). MIT Press is preparing an English-language edition.

In the 1980s Grothendieck had expected the work to be published by Éditions Christian Bourgois. He had even written an introductory foreword. In the end, however, this publication never came about.

The esoteric nature of the text doesn't fully explain why such a major work remained unpublished for so long. A more direct explanation lies in the unfounded accusations contained in the manuscript, which Grothendieck refused to edit out. Notably, he accused his students of having abandoned his work and "buried" him, which is absurd (on this subject see Serre's spot-on response in his letter to Grothendieck of July 23, 1985). Other accusations were outright defamatory and would have exposed the publisher to charges of libel.

During the 2000s, a collective called the Grothendieck Circle worked to edit and make openly accessible many other unpublished texts and documents, *Harvests and Sowings,* for one, but also *La clef des songes* (The key to dreams), another remarkable yet deeply esoteric text.

This work was interrupted after Grothendieck circulated a "declaration of the intention not to publish" dated January 3, 2010, in which he affirmed the following: "I have no intention of publishing, or of re-publishing, any work or text, in whatever form, of which I am the author. . . . Any publication or dissemination of such texts which have appeared in the past without my consent, or which may appear in the future during my lifetime, against my express wishes detailed herein, is in my view illegal."

However, one month later, on February 3, 2010, Grothendieck reaffirmed the importance of *Harvests and Sowings* in a letter to the mathematician Frans Oort, cited in Ching-Li Chai and Frans Oort, "Life and Work of Alexander Grothendieck," *Notice ICCM* 5, no. 1 (2017): 22–50. It is also the source of this quote (originally in English): "This 'Reflection and Testimonial' on my life as a mathematician, unreadable as it is I admit, has much meaning for me, if not to anyone else!"

An excellent introduction to Grothendieck's biography is Allyn Jackson's two-part article, "*Comme Appelé du Néant*—As If Summoned from the Void: The Life of Alexandre Grothendieck," *Notices of the American Mathematical Society* 50, no. 4 (2004): 1038–56, and 51, no. 10 (2004): 1196–1212, online at https://www.ams.org/notices/200409/fea-grothendieck-part1.pdf; and https://www.ams.org/notices/200410/fea-grothendieck-part2.pdf.

Chapter 8. The Theory of Touch

The pages describing the theory of touch in terms of points and pits are written in a style that is reminiscent of actual mathematical research articles. If you liked this passage, you would probably also like official mathematics.

Thurston, "On Proof and Progress in Mathematics," *Bulletin of the American Mathematical Society*, no. 30 (1994): 161–77, https://arxiv.org/pdf/math/9404236.pdf.

Chapter 9. Something's Going on Here

In three dimensions, there are five convex regular polyhedra (the correct definition of *regular* is a bit technical, but it implies that all faces are identical regular polygons): the tetrahedron (four faces), the cube (six faces), the octahedron (eight faces), the dodecahedron (twelve faces), and the icosahedron (twenty faces). These five polyhedra have been known for millennia, and are notably mentioned in one of Plato's dialogues, *Timaeus*. Even though Plato simply reproduced a knowledge that had long preceded him, these five polyhedra have since become known as Platonic solids.

The notion of a regular polyhedron generalizes to any dimension; one then speaks of *regular polytopes*. These have been entirely classified, most notably thanks to the work of Ludwig Schläfli (1814–1895) and H. S. M. Coxeter (1907–2003). The classification leads to a very particular phenomenon in eight dimensions, with an exceptional and remarkable object called E_8, which will reappear in the notes to chapter 15.

"He's better than me." Grothendieck made this comment about Deligne in a private conversation with George Mostow (1923–2017), who repeated it to me personally.

Pierre Deligne's Abel Prize interview took the form of a conversation with two mathematicians, Martin Raussen and Christian Skau. The quotations are verbatim transcriptions of the live interview, available online at https://www.youtube.com/watch?v=MkNfooUt2TQ. An official transcription, which appeared in *Notices of the American Mathematical Society* in 2014, is accessible at https://www.ams.org/notices/201402/rnoti-p177.pdf.

Chapter 10. The Art of Seeing

Bill Thurston's childhood is recounted in David Gabai and Steve Kerckhoff, eds., "William P. Thurston, 1946–2012," *Notices of the American Mathematical Society* 62, no. 11 (December 2015): 1318–32, and 63, no. 1 (January 2016): 31–41, online at http://www.ams.org/notices/201511/rnoti-p1318.pdf and https://www.ams.org/notices/201601/rnoti-p31.pdf.

As for Thurston's geometric intuition, I highly recommend the animated film *Outside In,* adapted from one of his proofs and produced by the Geometry Center of the University of Minnesota, as well as the Landau Lectures, a series of classes given by Thurston in 1996 at the Hebrew University of Jerusalem. All of these videos are easily accessible online.

As regards color blindness ("Daltonism"): the frequency given of 8 percent for men is an estimation given for the population of northern Europe (https://en.wikipedia.org/wiki/Color_blindness). It stems from a coding error that blocks the expression of a protein and is thus a recessive mutation, and the gene is carried by chromosome X. From this and basic math, one can explain why the frequency among women is the square of that among men.

Dalton's original article, "Extraordinary Facts relating to the Vision of Colors," which appeared in 1798, indicates that the communication occurred on October 31, 1794. The article is remarkably well written and is perfectly readable even today.

"People don't understand how I can visualize in four or five dimensions."

Thurston's comments are reported in Leslie Kaufman, "William P. Thurston, Theoretical Mathematician, Dies at 65," *New York Times,* August 22, 2012.

The documentary *The Boy Who Sees without Eyes* (2007), directed by Elliot McCaffrey and available online, gives you an idea of the abilities of Ben Underwood. Studies of human echolocation suggest that, for the sightless, this faculty mobilizes regions of the brain which, for the sighted, deal with visual information (https://en.wikipedia.org/wiki/Human_ echolocation).

Chapter 11. The Ball and the Bat

The quotations from Daniel Kahneman are taken from *Thinking, Fast and Slow* (New York: Farrar, Straus and Giroux, 2011).

Chapter 12. There Are No Tricks

The anecdote about Bill Thurston is related in the biographical section in David Gabai and Steve Kerckhoff, eds., "William P. Thurston, 1946–2012," *Notices of the American Mathematical Society* 62, no. 11 (December 2015): 1318–32, and 63, no. 1 (January 2016): 31–41, online at http://www.ams.org /notices/201511/rnoti-p1318.pdf and https://www.ams.org/notices/201601/rnoti -p31.pdf.

Chapter 13. Looking Like a Fool

The quotations from Pierre Deligne are taken from the 2014 Abel Prize interview with Martin Raussen and Christian Skau: available online at https://www.youtube.com/watch?v=MkNfooUt2TQ or in *Notices of the American Mathematical Society,* accessible at https://www.ams.org/notices/201402 /rnoti-p177.pdf.

Chapter 14. A Martial Art

"Like an elephant or a panther": the quotations are taken from a letter from Descartes to Pierre Chanut dated March 3, 1649, in *Oeuvres de Descartes,* ed. Charles Adam and Paul Tannery (Léopold Cerf, 1897–1913), available at Wikisource, https://fr.wikisource.org/wiki. Chanut was not only the French ambassador to Sweden but a close friend of Descartes's.

Descartes wrote *Discourse on Method* in French (the full original title is *Discours de la méthode pour bien conduire sa raison, et chercher la vérité dans les sciences*). Choosing the vernacular, not Latin, was a major departure from the scholarly norms of his time. Descartes explained that he wanted his message to reach a wide audience, far beyond the (then all-male) world of academia and theology, and be "understood by women and children."

The quotations are from the English translation by Ian Maclean (Oxford: Oxford University Press, 2006.)

Descartes's recollection of his three dreams appeared in a text that has since been lost, *Olympica,* which is known only from a transcription made by Adrien Baillet (1649–1706), his first biographer, in *La vie de Monsieur Descartes* (1691). Baillet had access to many original documents and direct statements, and his text remains the unique reference for many aspects of the life and work of Descartes, including the eventful night of November 10–11, 1619.

All quotations and details regarding the dreams rely on Baillet's unverifiable paraphrase of *Olympica.* Baillet is also our only source for the information regarding *L'art d'escrime* (The art of fencing).

Echoing the themes of chapter 6, one also finds in Baillet's book this interesting side comment about Descartes: "It must nevertheless be admitted that he did not read a great deal, that he had few books, and that most of them found among his possessions at his death were gifts from friends."

Rules for the Direction of the Mind was written in Latin (*Regulae ad Directionem Ingenii*). The quotations are from the English translation by John Cottingham, Robert Stoothoff, and Dugald Murdoch, *The Philosophical Writings of Descartes* (Cambridge: Cambridge University Press, 1985).

Chapter 15. Awe and Magic

The anecdotes and quotations about Cantor are taken from https://en.wikipedia.org/wiki/Georg_Cantor.

Here is a rough outline of how one can build an actual "proof" that a trefoil and an unknot are truly different. A general strategy to prove that two knots are different is to identify a "knot invariant" that differentiates them.

An invariant of a knot is a common characteristic shared among all its drawings (also called *diagrams*), however complicated they may be.

"Tricolorability" is a simple knot invariant that does the job here. By definition, a knot diagram is said to be *tricolorable* if you can color each "piece" of the diagram (meaning the visually apparent pieces of the diagram, obtained by considering that a line passing beneath another one is "cut" in two at the intersection) by using three distinct colors and respecting these rules:

—Each piece must be entirely painted with a single color.
—All three colors must be used, each at least once.
—At each intersection, the three pieces involved in the intersection (that which is "on top" and the two "beneath") must be either of three distinct colors or all of the same color.

Even though it's not obvious at first glance, you can prove that tricolorability really is a knot invariant (that is, whether a knot diagram is tricolorable depends only on the knot and not on the actual diagram). To prove this, one first uses a theorem of Kurt Reidemeister (1893–1971) that states that two diagrams represent the same knot if and only if you can go from the first diagram to the second by a series of basic transformations, the *Reidemeister moves* (proving this theorem is a bit technical). One then observes that the Reidemeister moves preserve tricolorability (this part is easier.)

For example, a trefoil knot is tricolorable, whereas an unknot is obviously not (it consists of a single piece and thus you can't use three different colors).

If a trefoil knot were the same as an unknot, they would either both be tricolorable or neither would. Thus, by exhibiting a knot invariant that differentiates them, you prove that they are two distinct knots.

Even if this allows you to prove the result, the definition of tricolorability seems as arbitrary as the famous "trick" for calculating the sum of whole numbers from 1 to 100.

As always, this apparent "trick" is the sign that there is a more profound

way to understand what is happening. I am unfortunately personally unable to explain it simply. It involves a kind of intuition that is difficult to communicate in just a few words.

As regards complicated diagrams of unknots, there is an interesting discussion thread started on the site MathOverflow by Timothy Gowers, Fields medalist 1998 ("Are There Any Very Hard Unknots?" https://mathoverflow.net/questions/53471/are-there-any-very-hard-unknots).

The "complicated" diagram of an unknot shown in this chapter is the so-called "gordian knot." It's from the German mathematician Wolfgang Haken (1928–2022), best known from having proved, with Kenneth Appel, the famous "four colors theorem."

A short YouTube video entitled *Haken's Gordian Knot Animation* illustrates why this diagram represents an unknot (https://www.youtube.com/watch?v=hznI5HXpPfE).

About Kepler's conjecture: even though the proof by Tom Hales requires a phenomenal amount of computer calculations, it also contains a profound and original "conceptual" component. It's not at all evident a priori that the conjecture could be reduced to a *finite* number of calculations and that these calculations *in practice* could be done on a computer.

Proofs making use of computers are a topic of debate among the math community: if no human can read and understand them, should we really consider them as being proofs? And how can we check that the software is itself correct?

Following his initial proof, Tom Hales started the ambitious project of "formalizing" his proof, that is, in producing a computer-based proof that could verify its own validity. Hales's attempt met with success. It's explained in a research presentation entitled *Formalizing the Proof of the Kepler Conjecture* available on YouTube at https://www.youtube.com/watch?v=DJx8bFQbHsA. This presentation, given by Hales in Paris in 2014, wasn't meant for a general audience, which makes it even more interesting for outsiders, as an opportunity to get a feel for the "living" reality of contemporary math research.

As regards dimensions 8 and 24, an explanation of why it is possible to determine the densest sphere packings lies in the existence of exceptional geometric structures that are specific to these dimensions and give rise to unusually dense packings. The methods used by Maryna Viazovska rely on these particular phenomena and are specific to these dimensions.

In dimension 8 the exceptional structure is called E_8 (see the notes for chapter 9 on the classification of polytopes). In the associated sphere packing, the number of contacts between neighboring spheres (what is called the *kissing number*) is 240.

In dimension 24, the geometry of the pile is that of the "Leech lattice," an exceptional structure specific to dimension 24 (https://en.wikipedia.org/wiki/Leech_lattice). The kissing number of 196,560 evokes the dimension 196,883 mentioned in chapter 20 and associated with the "Monster." It's not a coincidence. Mathematicians know that these kinds of numerological oddities are often the sign of much more profound connections. The Monster, one of the most intriguing math objects, is connected to a number of other exceptional structures, most famously through the "monstrous moonshine" (see https://en.wikipedia.org/wiki/Monstrous_moonshine).

Chapter 17. Controlling the Universe

Concerning the Unabomber, apart from the Wikipedia page, which is quite extensive, our principal sources are:

—On Ted Kaczynski's childhood, the televised interview with his brother David Kaczynski: https://www.youtube.com/watch?v= K20H5pFWEjo.
—On the extracts from his diaries: David Johnston, "In Unabomber's

Own Words: A Chilling Account of Murder," *New York Times,* April 29, 1998.

—On the attack on American Airlines Flight 444: Stephen J. Lynton and Mike Sager, "Bomb Jolts Jet," *Washington Post,* November 16, 1979.

—The Unabomber's manifesto is available online, for example, at the *Washington Post* website: https://www.washingtonpost.com/wp-srv/national/longterm/unabomber/manifesto.text.htm.

—On the details concerning the attacks and the investigation, see the conference given on November 19, 2014, at the District Court of Sacramento, filmed and shown on C-SPAN: https://www.c-span.org/video/?322849–1/unabomber-investigation-trial.

—Bill Thurston's role in the investigation is related by Steven G. Krantz in *Mathematical Apocrypha: Stories and Anecdotes of Mathematicians and the Mathematical* (Mathematical Association of America, 2002).

The first quotation (undoubtedly false) attributed to Grisha Perelman ("What would I do with a million dollars when I can already control the universe?") was reported by a "journalist and producer" who claimed to be a close friend of Perelman's who was preparing a documentary on him. The quotation was repeated in the Russian tabloid *Komsomolskaïa Pravda.* The documentary was never produced and the source is doubtful.

The second of Perelman's quotes ("Money and fame don't interest me") comes from the article "Russian Maths Genius Perelman Urged to Take $1m Prize," *BBC News,* March 24, 2010, http://news.bbc.co.uk/2/hi/europe/8585407.stm.

The discussion on the collaborative site MathOverflow between Bill Thurston and Muad is available online at https://mathoverflow.net/questions/43690/whats-a-mathematician-to-do. It should be noted that Thurston's comment "I try to write what seems real. By now, I have no cause to fear how I will be judged, which makes it much easier for me" speaks to the theme taken up in the epilogue: in order to recount the human experience of understanding, mathematicians need to overcome the reticence of their community to discuss subjects that "aren't serious." Even world-class mathematicians can be intimidated by Hardy's curse.

Chapter 18. The Elephant in the Room

The *species problem,* the impossibility of rigorously defining what constitutes an animal species, is a well-known epistemological problem that has been widely discussed.

The *paradox of the heap* (also known as the *sorites paradox* from the Greek word for heap) is among the other classic problems that show the fuzziness of human language. If you take a grain of sand from a heap, it remains a heap of sand, but if you continue, at some point it ceases to be a heap. But at what point is the limit? This problem, like that of the bald man (if you take one hair from the head of a man, that doesn't make him bald, but can you really define the border between being bald and not being bald?), is generally attributed to Eubulides, a Greek philosopher of the fourth century BCE.

The quotations from Ludwig Wittgenstein are taken from paragraphs 106 and 107 of *Philosophical Investigations,* 4th ed. (Oxford: Wiley-Blackwell, 2009), a text completed in 1949 and published posthumously in 1953. At the beginning of his career, however, Wittgenstein seemed close to the logicist position of Bertrand Russell (see my epilogue), as opposed to the preoccupations of *Philosophical Investigations.* The later works of Wittgenstein are an excellent complement to this chapter as well as to chapter 19. The most accessible is perhaps *On Certainty,* a short text assembled posthumously from his notes (Oxford: Wiley-Blackwell, 1975).

Chapter 19. Abstract and Vague

The question of the nature of abstract concepts is known in philosophy as "the problem of universals." The "realist" position holds that the concepts are "real" things, that they exist independently of human cognition. The "nominalist" position (and its "conceptualist" variant) holds that they are verbal constructs (or things that exist only in our heads). Historically, the realist position was dominant. During the Middle Ages in Europe the question was the object of a heated debate known as "the quarrel of universals," stoked notably by the positions taken by Pierre Abélard (1079–1142) and William of Ockham (1285–1347), whose conceptualist stances were condemned by the Church. In a sense, deep learning has vindicated them.

As for the specialization of neurons, a famous article that appeared in *Nature* describes the observation of a "Jennifer Aniston neuron" that reacts specifically to the presence of the actress in an image. See R. Quian Quiroga, L. Reddy, G. Kreiman, et al., "Invariant Visual Representation by Single Neurons in the Human Brain," *Nature,* no. 435 (2005): 1102–7.

Chapter 20. A Mathematical Awakening

The two quotations from Bill Thurston are taken from his preface to *The Best Writing on Mathematics, 2010,* ed. Mircea Pitici (Princeton, NJ: Princeton University Press, 2011).

The quotations from Grothendieck are from *Harvests and Sowings.*

The article by Bob Thomason and Tom Trobaugh is "Higher Algebraic K-Theory of Schemes and of Derived Categories," in *The Grothendieck Festschrift,* vol. 3 (Boston: Birkhäuser, 1990), 247–429.

Philosophy of mathematics has traditionally framed the debate on the "existence" of the parallel world of perfect mathematical entities as an opposition between Platonism (which posits that it does exist) and formalism (which posits that mathematics is nothing more than ink on paper, subject to a typographical game of mechanical deduction).

In "Some Proposals for Reviving the Philosophy of Mathematics," *Advances in Mathematics* 31 (1979): 31–50, Reuben Hersh (1927–2020) rightly points out that "the typical 'working mathematician' is a Platonist on weekdays and a formalist on Sundays."

Indeed, you can't do any work without projecting yourself into a fantasy world where mathematical objects "truly exist," but when your in-laws become inquisitive about your job, it's safer to characterize it as anchored in computer-like calculations.

Hersh states that both positions are equally untenable and the whole debate is entirely pointless: "The alternative of Platonism and formalism comes from the attempt to root mathematics in some nonhuman reality. If we give up the obligation to establish mathematics as a source of indubitable truths, we can accept its nature as a certain kind of human mental activity."

Hersh's brilliant 1979 article is the earliest instance I could find in the literature where the notion that *mathematics is best defined as a particular human activity* is articulated in full clarity, decades ahead of Thurston.

Epilogue

In his first letter to G. H. Hardy, dated January 16, 1913, Srinivasa Ramanujan claimed he was twenty-three years old. Born in December 1887, he would actually have been twenty-five. I can find no explanation for this discrepancy.

Hardy's review of *Principia Mathematica* is "The New Symbolic Logic," *Times Literary Supplement,* September 7, 1911.

It was to Hardy that Russell recounted having had the following nightmare: in the distant future, there remains only a single copy of *Principia Mathematica,* kept in a large university library. An employee of the library is charged with searching the shelves for books that have become useless in order to free up more space. The employee picks up the last copy of *Principia Mathematica* and hesitates. (This anecdote is reported by Hardy in *A Mathematician's Apology* [1940; repr., Cambridge: Cambridge University Press, 1992].)

Beyond its inhuman character, the formalist project that underpins *Principia Mathematica* is also problematic from a logical perspective. Kurt Gödel (1906–1978) famously proved with his 1931 incompleteness theorems that formal systems such as those in *Principia Mathematica* always contain "undecidable" statements (that is, statements that are not provable, and whose negation is also not provable.)

As a first-year student at the École normale supérieure, I followed a course given by Xavier Viennot in which he introduced us to peculiar "intuitive" objects, akin to Lego or Tetris blocks, that we could mentally manipulate to find "visually evident" some of Ramanujan's results. This mind-blowing experience changed my entire perception of mathematics: I realized that formulas as esoteric as those of Ramanujan could in fact transcribe nonverbal intuitions that were both simple and subtle. A good illustration of his approach is the presentation he gave in Chennai (formerly Madras) in 2019: "Proofs without Words: The Example of Ramanujan Continued Fractions"; notes: http://www.xavierviennot.org/coursIMSc2017/lectures_files/Ramanujan

Inst_2017.pdf; video recording: https://www.youtube.com/watch?v=jQchTFnKBQs.

Other works cited in this chapter:

Alfred North Whitehead and Bertrand Russell, *Principia Mathematica,* vol. 1 (Cambridge: Cambridge University Press, 1910).

Misha Gromov, "Math Currents in the Brain," in *Simplicity: Ideals of Practice in Mathematics and the Arts,* ed. R. Kossak and P. Ording (Cham: Springer, 2017).

Hardy, *A Mathematician's Apology.*

Illustration Credits

Line drawings and charts courtesy of Éléonore Lamoglia.

Literate and illiterate world population among people aged 15 and older by Our World in Data based on OECD and UNESCO (2016).

Photograph of the dolphin Wave performing a tailwalk courtesy of Dr. Michael Bossley.

Photograph of the Fosbury high jump copyright Raymond Depardon / Magnum PhotosTEST 01-Stock (Photogs).

Photograph of Alexander Grothendieck by Konrad Jacobs, CC BY-SA 2.0 DE.

Photograph of Bill Thurston as a child courtesy of Rachel Findley.

Photograph of Thomas Hales by Bob Kalmbach, ID 2b0c0b4268, Michigan Photography digital collection, Bentley Historical Library, University of Michigan.

Photograph of young Ted Kaczynski by George M. Bergman, Oberwolfach Photo Collection, GFDL.

Diagram of a neuron by Nicolas Rougier, CC BY-SA 4.0.

Illustration Credits

Acknowledgments

My deepest thanks go to Hélène François, my first reader, for her generosity, heart, and visionary spirit. This book owes much to her.

My friends Farouk Boucekkine, Michel Broué, Nicolas Cohen, Hélène Devynck, Marion Gouget, Basile Panurgias, and Jérôme Soubiran have been my test readers and sparring partners throughout the initial writing phase. I also owe a lot to them.

This first English-language edition is much more than a translation. English has long been my exclusive language for writing mathematics, and I initially thought that it would be the best language for me to write *about* mathematics. However, I wanted this book to be emotionally accurate. It soon became obvious that I would never achieve this without switching to French, the language of the *child within me*. Ironically, still, many passages spontaneously came to me in English, and I had to mentally translate them into French.

My secret plan was for the first English version to be a full-fledged second edition. I am extremely grateful to Jean Thomson Black and Yale University Press for allowing this to happen, and for sparing no efforts.

Working with Kevin Frey was a pleasure. His initial translation was already in tune with my own internal voice, which made the revision effort natural and easy for me. This process took place directly in English, with Kevin and I working hand in hand. There are no major changes compared to the original edition, but a few sections have been rewritten and clarified, and there are many small tweaks at the sentence and paragraph levels. I thank Kevin for his patience and curiosity. I also thank the three anonymous reviewers who provided detailed feedback on this edition, and the countless readers who reached out with comments on the original edition and helped me understand where it could be improved.

I thank the whole team at Yale University Press, especially Elizabeth Sylvia and Joyce Ippolito. Thanks to Robin DuBlanc for the copyediting.

When my original editor, Mireille Paolini, first read the manuscript, she commented that it was an American book written in French. I hope that she'll like this version. I thank Mireille for her absolute confidence and relentless commitment. Thanks to the entire team at Éditions du Seuil, in particular Adrien Bosc, Hugues Jallon, Séverine Nikel, Emmanuelle Bigot, Muriel Brami, Bénédicte Gerber, Joséphine Gross, Virginie Perrollaz, and of course Maria Vlachou and the international rights team. Thanks to Éléonore Lamoglia, who produced the beautiful illustrations.

Thanks to Ardavan Beigui, Fabrice Bertrand, Simon Boissinot, Emmanuel Breuillard, Olivia Custer, Lucas Dernov, Maxime Dernov, Nicolas François, Artem Kozhevnikov, Vincent Levy, François Loeser, Raphaël Rouquier, Vincent Schächter, Claudia Senik, Marguerite Soubiran, Sarah Stern, Solal Stern, Maxime Verner, and Agathe Vernin, whose careful reading and comments have made this a much better book.

For their help with documentation and permissions, thanks to Sophie Kucoyanis and Éditions Gallimard, MIT Press, Mike Bossley, Steve Krantz, Laurent Fleury, David Gabai, Steven Kerckhoff, and Rachel Findley.

I thank the people who taught me how to think.

Index

Note: Page numbers in italics indicate a figure.